AFGHANISTAN
AUSTRIA
BAHRAIN
BERMUDA
CHINA
CUBA
EGYPT
ETHIOPIA
REPUBLIC OF GEORGIA
GERMANY
KUWAIT
IRAN
IRAQ
ISRAEL
MEXICO
NEW ZEALAND
PAKISTAN
RUSSIA
SAUDI ARABIA
SCOTLAND
SOUTH KOREA
UKRAINE

# Iraq

Angelia L. Mance
Jacksonville State University

Series Consulting Editor
Charles F. Gritzner
South Dakota State University

*Frontispiece:* Flag of Iraq

*Cover:* Donkeys carry the harvest, 1990.

**CHELSEA HOUSE PUBLISHERS**

VP, NEW PRODUCT DEVELOPMENT  Sally Cheney
DIRECTOR OF PRODUCTION  Kim Shinners
CREATIVE MANAGER  Takeshi Takahashi
MANUFACTURING MANAGER  Diann Grasse

**Staff for IRAQ**

EDITOR  Lee Marcott
PRODUCTION EDITOR  Jaimie Winkler
PICTURE RESEARCHER  Pat Holl
COVER AND SERIES DESIGNER  Takeshi Takahashi
LAYOUT  21st Century Publishing and Communications, Inc.

©2003 by Chelsea House Publishers, a subsidiary of Haights Cross Communications. All rights reserved. Printed and bound in the United States of America.

A Haights Cross Communications Company

http://www.chelseahouse.com

First Printing

1 3 5 7 9 8 6 4 2

Library of Congress Cataloging-in-Publication Data

Mance, Angelia L.
  Iraq / Angelia L. Mance.
    p. cm. — (Modern world nations)
Includes index.
Summary: Describes the history, geography, government, economy, people, and culture of Iraq.
  ISBN 0-7910-6928-1
  1. Iraq—Juvenile literature. [1. Iraq.] I. Title. II. Series.
DS70.62 .M36 2002
956.7—dc21

2002007325

# Table of Contents

| | | |
|---|---|---|
| 1 | Introducing Iraq | 9 |
| 2 | Natural Landscapes | 17 |
| 3 | Iraq Through Time | 29 |
| 4 | People and Culture | 51 |
| 5 | Government | 61 |
| 6 | Economy | 69 |
| 7 | Living in Iraq Today | 77 |
| 8 | Iraq Looks Ahead | 83 |
| | Facts at a Glance | 88 |
| | History at a Glance | 90 |
| | Further Reading | 92 |
| | Bibliography | 93 |
| | Index | 94 |

# Iraq

View of Baghdad.

# Introducing Iraq

Iraq lies in a region that is called the "Dry World." The region is a vast desert that stretches across North Africa and reaches on into central Asia. But most of the people live in a region near water—in the broad Tigris-Euphrates basin. This is an area of the world where water is almost always at a premium, where peasants often struggle to make soil and moisture yield a small harvest, and where in the future water may be a greater source of conflict than oil.

## Origins

The region that is home to Iraq is also referred to as the "Arab World," although this term implies a sameness that does not in truth exist. The name Arab is given inaccurately to people who speak Arabic and Arabic-related languages, but ethnologists (scientists who study racial origins and cultures) restrict it to certain occupants of the Arabian Peninsula—the Arab "source." Another name given to the region is the "Islamic World." This region was the birthplace of the prophet Muhammad in A.D. 571. In the centuries after Muhammad's death in 632, Islam spread into Africa, Asia, and Europe by means of Islamic conquest and expansion. Armies penetrated southern Europe, their caravans crossed the deserts, and their ships traded along the coasts of Asia and Africa.

Along the routes of trade and conquest, armies carried the Muslim (Islamic) faith, converting the ruling classes in western Africa, challenging the dominance of Christianity in the highlands of Ethiopia, advancing into the deserts of inner Asia, and making headway into India and throughout Southeast Asia. Islam was the religion of the merchant and the military, and these travelers carried it with them wherever they journeyed. The more than 1 billion followers of the Islamic faith are proof of this vast spread, which extends well beyond the limits of what we know as the Islamic World. Thus, the name, World of Islam, as applied to this region, is not entirely appropriate either. The Islamic religion extends far beyond the areas it is most known for, and within the region there is less uniformity than the name implies.

The area centering around Iraq is also known as the Middle East. This name reflects the bias of the Western world, which sees a country such as Iraq as the Middle East and countries such as China and Japan as the Far East. Geographically, the region can be called Northern Africa/Southwest Asia.

The Republic of Iraq is a country in southwestern Asia that has been home to some of the world's greatest civilizations. For example, the ancient civilizations of Assyria, Babylon, and Sumer developed in what we know today as Iraq. The modern state of Iraq was created in 1920 by the British government, whose forces had occupied it during World War I (1914–1918). Baghdad, Iraq's capital, is also the country's largest city.

Iraq has a population of 23.6 million, roughly two-thirds the population of California. Because of its major oil reserves and large areas of irrigated farmland, Iraq is probably the best endowed with natural resources when compared with its neighbors. Descended from the early Mesopotamian states and empires that emerged in the basin of the Tigris and Euphrates rivers, Iraq also has a rich history. Its archaeological sites rank among the world's most important.

## Conflicts with Neighboring Countries

Iraq is bounded by six neighboring countries and has had resentments and conflicts with most of them in recent years. Turkey, the source of both of Iraq's vital rivers, lies to the north. To the east lies Iran, with whom Iraq engaged in a decade-long devastating war during the 1980s. At the head of the Persian Gulf lies Kuwait, which was invaded by Iraq's army in 1990. To the south lies Saudi Arabia, an ally of Iraq's rivals. And to the west, Jordan and Syria adjoin Iraq. Jordan provided Iraq with an outlet to the sea through the port of Aqaba during the 1991 Gulf War in defiance of United Nations (UN) sanctions (rules). However, relations between Jordan and Iraq have since become worse. Syria lies farthest from Iraq, despite its shared long desert border. The Euphrates River passes through Syria before it enters Iraq. Basically, water is always a potential source of great conflict in this arid zone.

Iraq's southern zone has great significance because the Tigris and Euphrates rivers, lifelines of the country, join to become the Shatt-al-Arab, Iraq's water outlet to the Persian Gulf. Over its last 50 miles (80 km) or so, the Shatt-al-Arab waterway also becomes the boundary between Iraq and Iran. Conflict over Iraq's claim to the land on the Iranian side of the waterway helped to bring about the war that began between Iraq and Iran in 1980. Situated at the northern tip of the Persian Gulf, Iraq's coastline is only about 19 miles long. Its only port on the gulf, Umm Qasr, is small and shallow.

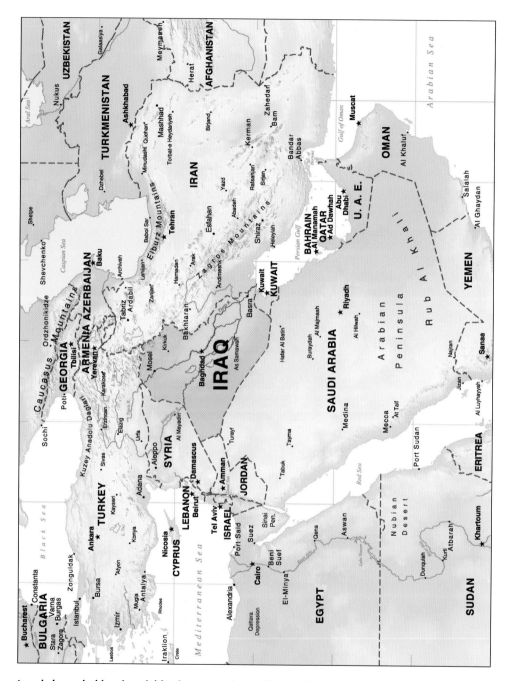

Iraq is bounded by six neighboring countries and has had resentments and conflicts with most of them in recent years. Turkey, Iran, Kuwait, Saudi Arabia, Jordan, and Syria all adjoin Iraq.

Iraq's most important southern city, Basra, and the position of the boundary with Kuwait are also important. This is because several major oil fields are found here. One, the Rumailah oilfield, extends from beneath Iraqi soil into Kuwait. In 1990, Iraq claimed that Kuwait was drilling oil from this field by using slanted wells under Iraqi territory. Iraq also claimed that the boundary between the two countries was never agreed upon and that Kuwait was not adhering to OPEC (Organization of Petroleum Exporting Countries) rules on oil production and pricing. The Shatt-al-Arab waterway was still filled with the wreckage of the just-ended Iran-Iraq War, so that the annexation of Kuwait would also give Iraq a new outlet to the Persian Gulf. These justifications, combined with the potential returns, persuaded Iraq to embark on its unsuccessful invasion of Kuwait in 1990.

**The Population**

Most of Iraq's population is Arab, and the country has been politically active in the Arab world since it became a modern state in 1920. Most of its political regimes have supported either total Arab policies, or, at the very least, partial Arab unification. Iraq became a republic in 1958, but has been a dictatorship dominated by a single party, the Ba'th Party, since 1968. Saddam Hussein has led the country through his dictatorship since 1979. Throughout Hussein's presidency, Iraq's national and foreign policies have been ambitious and have often involved great risk. Hussein's failures have been high profile—the war with Iran, the invasion of Kuwait, and attempts to topple other Arab regimes.

The core area of Iraq, centered in Baghdad and the Shi'a Muslim-dominated south, are two of Iraq's major subregions. A third subregion lies in the north—the land of the Kurds. Although fewer than 4 million Kurds live in the mountainous area of northern Iraq, they constitute as much as 15 percent of the total population in the region. Actually, the Kurds are minorities in all the countries in which they reside. Estimates place the number of Kurds found in the region, which includes Turkey, Iran, and Iraq, at 25 million.

Iraq's Kurds occupy an important area of the country, since the huge oil reserves on which the city of Kirkuk is situated lie in Kurdish territory. During both the Iran-Iraq and Gulf wars, the Kurds rose against their Iraqi rulers and briefly succeeded in controlling parts of their homeland. But Iraq's response always ended such efforts, often with great loss of life. Saddam Hussein used cyanide and mustard gas against the Kurds in the 1980s. At the end of the Gulf War, the UN established a security zone between the 36th parallel and Iraq's northern border to encourage refugees who had crossed into Turkey and Iran to return. Much of the Kurdish territory, however, lies outside the safety zone. Throughout their history, the Kurds have always been helpless victims at the hands of their powerful neighbors.

**Present and Future Turmoil**

Although Iraq's infrastructure (e.g., bridges, roads, and power supplies) and economy were shattered during the Gulf War, the country already had wasted much of its potential on the earlier conflict with Iran—as well as on mismanagement, corruption, and inefficiency. Iraq is blessed with good land for farming and a vast oil income and so should be one of the best economic success stories of the entire realm. Instead, its botched leadership has made it one of the world's tragedies. Apart from the problems brought about by the two wars in as many decades, Saddam Hussein has not complied with allowing weapons inspectors from the UN into the country to dismantle its weapons of mass destruction. This has brought about economic sanctions that have put the country into further despair. Still, Iraq has had some relief, both legal and illegal. Under the oil-for-food program, the UN allows Iraq to export unlimited quantities of oil in exchange for humanitarian relief supplies, such as food, medicine, and spare parts for

Introducing Iraq 15

An oil pump station in Iraq's largest oil center, located in Kirkuk in Kurdish territory.

vehicles. On the other hand, illegal trade in oil earned Iraq almost $2 billion in illegal income in the year 2000.

There may be more trouble ahead for Iraq. If one issue can cause the entire region (other than Islam and oil) to react, it is Israel. During the Gulf War, Iraq sought to draw Israel into the conflict by aiming missiles at the country. Although Iraq has tried in the past to build nuclear weapons to threaten Israel, it is the only Middle Eastern country without territorial contact with the Israelis. When the United States indicated in 2001 that Iraq might be a future target in the war on terrorism, Saddam Hussein once again threatened Israel. Indeed, Iraq faces more turmoil in the future under the oppressive regime of Saddam Hussein.

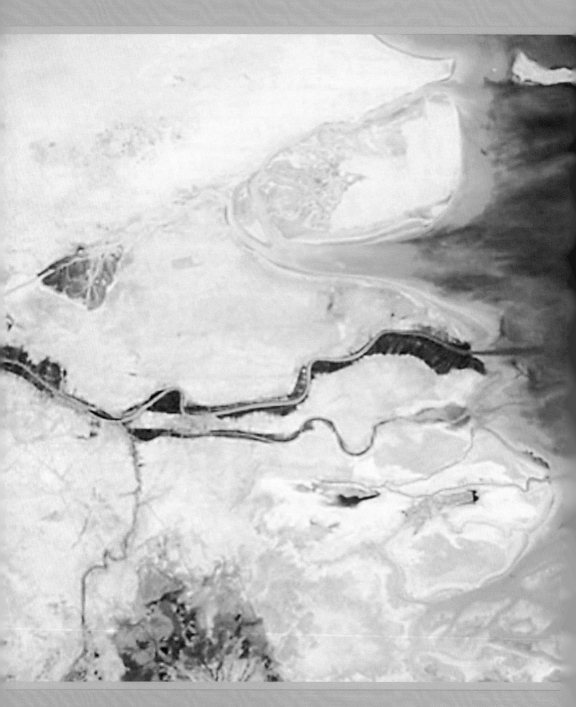
The Tigris-Euphrates delta as viewed from a satellite.

# CHAPTER 2

# Natural Landscapes

The ancient name for the territory of Iraq is *Mesopotamia*—"land between the rivers." The two rivers are the Tigris and the Euphrates. The modern name of Iraq is the Arabic word for cliff, a prominent geographic feature of the country.

With an area of 167,925 square miles (270,258 square kilometers), Iraq has four main geographic regions: desert in the west and southwest; dry, rolling grasslands between the upper Tigris and Euphrates rivers; highlands in the north and east; and a plain through which the lower Tigris and Euphrates flow.

## Physical Features

The Syrian Desert that stretches west and south of the Euphrates River is part of a larger desert area that extends into Syria, Jordan, and Saudi Arabia. The population is sparse in the parched desert areas, because there is little water. The desert plain is lined with *wadis*—streams that are dry most of the year. During the winter season, rain sends dangerous flash floods through these wadis. The desert areas of the west and southwest often develop thick, attractive plant cover in the spring after winter showers, but the vegetation soon gives way to barren land for the rest of the year.

The rolling grasslands between the Tigris River north of Samarra and the Euphrates River north of Hit are sometimes called al-Jazirah (the island). Here, the water has cut deep valleys into the land, which makes irrigation difficult. Still, rain-fed agriculture is possible in the region. The Tigris and Euphrates rivers are an invaluable resource, and this huge oasis undoubtedly helped to makes the beginning of civilization possible here. People were able to settle in this arid region, and the rivers made agriculture possible, with transport of food and supplies to market by boat.

**River Paths**

The Tigris River originates in Turkey before flowing into Iraq. The Euphrates also begins in Turkey, but flows through Syria before entering Iraq. The rivers meet at al-Qurnah in southern Iraq, forming the river known as the Shatt-al-Arab, which then flows into the Persian Gulf. This waterway forms a portion of the boundary between Iran and Iraq and was involved in the boundary dispute that led to war between the two countries in the 1980s.

The Tigris River is narrower than the Euphrates, but it carries more water. Both rivers begin to meander just south of the capital city of Baghdad, then flow through well-defined channels north of the city. The rivers have frequently changed course, leaving behind abandoned flood riverbeds, which allow spring floods to cover a large area. The Tigris can rise at the rate of 12 inches (30.5 centimeters) per hour, and flash floods are apt to occur during heavy rains. Systems of

# Natural Landscapes 19

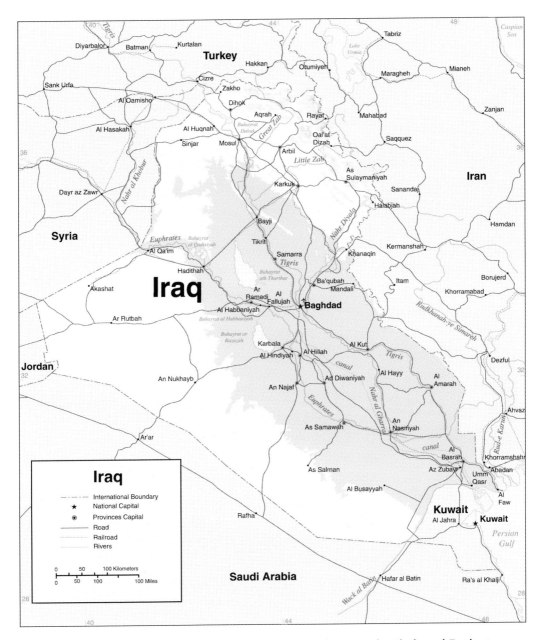

Iraq's southern zone has great significance because the Tigris and Euphrates Rivers, which are the lifelines of the country, join to become the Shatt-al-Arab, Iraq's outlet the Persian Gulf. The core of Iraq is centered in the city of Baghdad and in the Shia Muslim-dominated south, which are two of Iraq's major subregions. A third subregion lies in the north where the Kurds live.

flood control and irrigation are in place to help prevent the devastation that has occurred in the past.

The Tigris and Euphrates rivers have been used as water highways throughout their existence. Still, they create problems for modern transportation. The lower section of the rivers often floods, which hinders road building. And the fast currents of the upper sections prevent boats from traveling upstream. The rivers are wide and slow near the Persian Gulf, but are also very shallow. Dredging (use of an apparatus to deepen the water bed) has been necessary to allow shipping in these waters.

**Vegetation**

Vegetation in the rolling grasslands between the Tigris and Euphrates rivers is a mixture of scattered, low-growing, perennial shrubs that live throughout the year and survive the severe heat. In the spring, bright flowering annuals abound after receiving winter rains.

Just southwest of the cities of Mosul and Kirkuk, the highlands of Iraq begin and extend to the borders with Iran. The foothills and plains meet mountain ranges from 3,000 to 12,000 feet (914 to 3,657 meters) near the borders of Iran and Turkey. In this mountainous area, there are several valleys suitable for agriculture, as the majority of Iraq's rainfall occurs in the northeast. The foothills and plains that receive so little rain cannot support diversified crops. The steady heat and sandy soil of the southeast allow for good quantities of cotton and a large date crop to be produced.

Vast stands of oak once covered the northern Kurdish mountains. These trees have disappeared over the years because they have been cut down for use as firewood and charcoal. Some areas are populated by other species of trees such as the pine and maple, whereas other areas appear to be little more than scrubland. Unfortunately, no forestry policies exist to prevent the cutting of the remaining trees.

Throughout the plains, common vegetation includes rockroses, sedges, capers, boxthorns, and rough grasses. In the spring daisies, buttercups, and poppies thrive. In the cities, date palms, orange and pear trees, rubber plants, and eucalyptus are common. Thickets of tamarisk and bulrushes grow along the riverbanks and marshlands, as well as poplar and willow trees.

**The Plain**

The low plain of Baghdad begins north of the city and extends to the Persian Gulf. Most of Iraq's people live on or near this plain. The plain is alluvial (composed of clay, sand, or other loose material)—built up by the mud and sediment laid down by the rivers. The Tigris and Euphrates rivers are above the level of the plain in many places—as is the Mississippi River in the southern United States.

This 45,000-square mile (72,423-square kilometers) area of plain is known as the delta, the point at which rivers flow into the gulf—an area grooved by river channels and irrigation canals. Temporary lakes form when the rivers flood, and the silt carried by the river, the irrigation canals, and the wind build up the delta plains at the rate of about eight inches every 100 years. Up to 12 inches (30.5 centimeters) of mud can be deposited in temporary lakes as the result of heavy flooding, and, unfortunately, the rivers also carry large quantities of salts. These salts concentrate near the surface because of the high water table and poor drainage, damaging crops and limiting successful farming.

A large part of the delta was once a marshland that extended into Iran. At the end of the Gulf War, Saddam Hussein built a canal to drain the marshes and divert the water to desert land. This was done in part to destroy rebel bases hidden in the marshland. Residents of the area, known as the Marsh Arabs, had to flee for their safety across the border to Iran.

Arabs living and working in marshes of Qurnah, Basra, Iraq in 1976.

## Weather and Climate

Iraq has two climatic provinces: the hot, arid lowlands, including the alluvial plains and the deserts; and the damper northeast, where the higher elevation produces cooler temperatures. In the northeast, rain-fed cultivation is possible, but elsewhere farmers must irrigate the land.

People of the lowlands experience two seasons, summer and winter, with a short transitional period between. In summer, which lasts from May to October, there are clear skies, extremely high temperatures, and low relative humidity. Very little, if any, rain falls from June through September. In Baghdad, July and August bring daily temperatures of about 95° F (35°C), and summer afternoon highs can reach 123° F (50°C). Temperatures range considerably between day and night in summer.

During winter, the paths of west-to-east-moving storm systems cross the Middle East, shift southward, and bring rain to southern Iraq. Annual amounts vary greatly from year to year, but mean annual rainfall in the lowlands ranges from four to seven inches; about 90 percent of the rainfall occurs between November and April. Winter in the lowlands lasts from December to February. Temperatures are generally mild, although extremes of hot and cold, including frosts, can occur. Winter temperature extremes in Baghdad range between 35° and 60° F (1°C and 15°C).

Summer in the northeast is shorter than in the lowlands, lasting from June to September, and the winter lasts longer. The summer is generally dry and hot, but average temperatures are about five to ten degrees cooler than those of lowland Iraq. Cold winters sometimes occur as a result of high relief and the influence of northeasterly winds that bring continental air from central Asia. January temperatures in Mosul range from 24° to 63° F (-4°C to 17°C), but temperatures as low as 12° F (-11°C) have been recorded.

Annual rainfall in the foothills of the northeast ranges from 12 to 22 inches (30.5 centimeters to 56 centimeters), enough to sustain a good seasonal pasture. Precipitation may exceed 40 inches (102 centimeters) in the mountains, much of which falls as winter snow. As in the lowlands, little rain falls during the summer.

The *shamal*, a steady northerly and northwesterly wind in summer, affects all of Iraq. It brings such dry air that hardly any clouds form, permitting intensive heating of the land surface by the sun. Another wind, the *sharki*, blows from the south and southeast during early summer and early winter; it is often accompanied by dust storms. Dust storms occur throughout Iraq during most of the year, and dust may rise to several thousand feet. More frequent in summer, five or six storms may strike central Iraq during July, the worst month.

The average annual rainfall is between 4 and 7 inches (10 and 18 centimeters) except in the mountains of the north and northeast, where rain is more plentiful. Because of the rugged terrain, however, mountain rain is not useful for the cultivation of crops. In the valleys, foothills, and plains, where 12 inches (30.5 centimeters) or more of rain falls each year, agriculture is possible without irrigation. But even here a shortage of rain can lead to crop failure, and only one crop a year can be grown.

Rain can be a problem as well as a gift in the desert areas. A few days of steady rain can turn roads into thick mud and disrupt mail and telephone service. Planted crops may be washed out. Houses and roofs made of adobe (clay) can leak badly. Fortunately, most buildings can be repaired when the rain stops. Soil and plants very quickly lose any moisture they get from the rain through evaporation.

## Soil

Two different types of soil are found in Iraq. Heavy alluvial deposits, containing a significant amount of humus (dark, organic soil) and clay, make up one soil type and are useful for construction. The lighter soils, lacking in humus and clay content, contain wind-deposited nutrients. A high saline (salt) content mars the otherwise rich composition of the soils. Irrigation and flood-control projects on the Tigris and

Natural Landscapes 25

Plowing arid land is difficult in the desert land of Iraq.

Euphrates rivers help to increase the agricultural production of this area.

## Environmental Issues

Serious environmental issues plague Iraq. Two destructive wars and years of economic isolation have seriously degraded Iraq's environment. The Iran-Iraq War (1980–1988) and the Persian Gulf War (1991) destroyed wildlife habitat, polluted Iraq's land and water, and led to the neglect of conservation efforts.

Much of Iraq's resources were destroyed during the Persian Gulf War, including equipment used in the petroleum industry. Although Iraq has destroyed many oil wells and refineries since the end of the war, the Iraqi government contends that the international economic embargo (restriction) established by the UN is preventing the repair of equipment needed to safely process the toxic by-products of oil refining. As a result, hazardous wastes are being released into the air or dumped into depleted wells.

## Wildlife

Birds are easily the most prominent wildlife in Iraq. About 390 species are found throughout the country. Some species of birds stay all year; many others are migrants and seasonal visitors, some of them attracted by the extensive wetlands. Species found in Iraq include crows, European robins, and storks. There are also vultures, buzzards, ravens, owls, and various species of hawks. Birds of prey such as the hawk are often trained for hunting. Other birds include ducks, geese, partridges, and sand grouse.

Among the most common mammals are deer, mountain goats, jackals, hyenas, wild boars, brown bears, rabbits, bats, wild cats, and squirrels. The lion has become extinct in Iraq, and the ostrich and the wild ass are nearly extinct. Rivers, streams, and lakes are well stocked with many kinds of fish, notably carp, barbus, catfish, and loach. Reptiles abound in the desert areas of Iraq, although lizards are far more numerous than snakes. In common with other regions of the Middle East, Iraq is a breeding ground for the unwelcome desert locust.

View of the ancient ruins of Ur.

# CHAPTER 3

# Iraq Through Time

## Early History

Once known as Mesopotamia, the land between the rivers, Iraq has seen a succession of early civilizations. Sumer was the earliest known civilization, founded around 4,000 B.C. and reaching its climax under the third dynasty of Ur around 3,000 B.C. With no natural barriers, the area succumbed to successions of foreign invaders, and the great civilizations of Babylon and Assyria followed. The Babylonian empire fell in 539 B.C. to Cyrus the Great of Persia (present-day Iran). The Persians retained control of the area for almost 200 years, until the conquest by the Macedonian king Alexander the Great in 331 B.C. Upon Alexander's death, the Greek Seleucid dynasty controlled Mesopotamia for about 200 years, which infused the region with Hellenistic (ancient Greek) culture. Persian rule followed the Greek Empire and lasted from about A.D. 150 until the Arab conquest swept through the region in the seventh century A.D.

The Arab-Islamic conquest of what is now Iraq started in 633 A.D and ended in 636 at the Battle of Qadisiyya, a village on the Euphrates River south of Baghdad. During this battle, an Islamic army defeated a Persian army that was six times larger than it. The Arab army moved quickly to the capital of the Sassanian Empire, where in 637 it seized the immense Persian treasury. Many tribes in the conquered lands were Christian Arabs. Some of them converted to Islam; others were allowed to stay, provided that they pay a tax.

From the mid-eighth century to 1258 A.D., Baghdad was the capital of the Abbasid caliphate or Islamic realm. Known as the Golden Age of Islamic power, the Abbasid period saw Baghdad become the second largest city in the known world after Constantinople and was the most important center of science and culture. The Abbasid realm was a forceful military power for a time, with its borders reaching southern France to the west and the China borders to the east. By the mid-ninth century, the Abbasid caliphate began a slow decline, as Turkic warrior slaves known as Mamluks became so prominent at the caliph's court that they almost monopolized power. In 945, the Buwayhids, Iranian Shi'a dynasty, conquered Baghdad. In 1055, the Seljuks, a Turkish Sunni clan, conquered the Buwayhids and reestablished Sunni rule in Baghdad. By the end of the 11th century, Seljuk power started to decline.

Hulagu, a grandson of the great Mongol conqueror Genghis Khan, took control of Baghdad in 1258. He killed all of the scholars and erected a pyramid from their skulls. The sophisticated irrigation system constructed by the Abbasids was destroyed. Iraq became a neglected frontier area ruled from the Mongol capital of Tabriz in Iran. In 1335, the last Mongol ruler of this region died. Chaos triumphed until the Turkic conqueror Tamerlane destroyed Baghdad in 1401. He massacred many of its inhabitants. Tamerlane's conquest marked the end of Baghdad's prominence.

Ottoman Turks and Iranian leaders competed for control of Iraq until the Ottoman Empire prevailed in the 17th century. The area came under Persian control in 1508, and by 1534 the Ottoman Turks conquered much of it. Persian armies took control of Baghdad and large parts of Iraq in 1623 until 1638, when Iraq was again brought under Ottoman rule. Iraq would remain under Ottoman control for almost three centuries.

**Modern History**

Modern Iraq's history began with the last era of Ottoman rule in the 1800s. Until the 1830s, Ottoman rule in Iraq was weak. Control of the area alternated between powerful local chieftains and Mamluk rulers. Some of the nomadic Arab tribes never really came under Ottoman control, and Kurdish dynasties maintained domination over the mountainous north. (Nomads are people who move from place to place, usually according to the seasons.) By the second half of the 18th century, the Mamluks had established effective control over the territory from Al Basrah to the north of Baghdad. The Mamluks, instituting central authority, introduced a functioning government. In 1831, the province of Iraq was divided into three administrative districts: Mosul, Baghdad, and Al Basrah. These districts were under the direction of the Ottoman administration. A series of governors held power briefly over the districts from 1831 to 1869, when Midhat Pasha, a distinguished Ottoman official, was able to impose effective central control throughout the region. Pasha is credited with modernizing Baghdad, in everything from transportation, the educational system, and basic needs such as sanitation. He imposed this authority on the tribal countryside, and for the first time the Arabs felt the impact of a strong Ottoman government, especially with regard to tax collections. The people resented this domineering central authority, and this resentment gave rise to a strong sense of Arab nationalism.

**Britain and Iraq**

Britain and Germany became rivals in the commercial development of the Mesopotamian area by the end of the 19th century. First interested in Iraq as a direct overland route to India, the British by 1861 had established a steamship company for navigating the Tigris to the port of Al Basrah. The route to India was critically important to Britain, because the British needed India's resources to fuel the vast British Empire. Germany had strong interests in the region as well, and announced plans to build a railroad from Berlin to Baghdad. British opposition to the project was overcome, and permission was granted from the Ottoman government to build a railroad from Baghdad to the Persian Gulf. To overcome this defeat, the British government managed to strengthen its position in the Persian Gulf area by establishing treaties of protection with local Arab chieftains. In 1901, British financiers were victorious in obtaining permission to explore and develop the oil fields of Iran. By 1909, the Anglo-Persian Oil Company (later known as the Anglo-Iranian Oil Company) was founded to support this new industry.

As the Ottoman Empire entered World War I (1914–1918) as an ally of Germany and Austria-Hungary, a British army division landed at Al Faw in southern Iraq and occupied Al Basrah. The landing filled the urgent British need to defend the Anglo-Persian Oil Company's interests in Iran. Steadily, the British army advanced northward against strong Ottoman opposition and entered Baghdad in March 1917. The British and the Ottoman Turks signed an armistice agreement in October 1918, although the British army continued to advance north, capturing Mosul in early November. With this victory, Britain extended its rule over almost all of Iraq.

In the early stages of World War I, the British government

King Faisal I, the first king of Iraq (about 1930).

was concerned with ensuring the interest of the Arabs in a military uprising against the Ottoman Turks. Arab leaders were promised that their people would receive independence if a revolt proved successful. By the summer of 1916, there was an uprising in Al Hijaz (the Hejaz), which was led by Faisal al-Husein, later to become Faisal I, the first king of Iraq. British General Edmund Allen provided skilled leadership and was aided by the tactical direction of British Colonel T. E. Lawrence.

Together, the Arab and British forces achieved great success in battles against the Ottoman army and were able to liberate much Arabian territory. After signing the armistice with the Ottoman government in 1918, the British and French governments issued a joint statement regarding their intention to help establish independent Arab nations in the Arab regions that were formerly under the control of the Ottoman Empire.

In 1919 at the Paris Peace Conference, the Allies (a coalition of the victorious nations in World War I) established Iraq (the territory encompassing the three former Ottoman districts of Mosul, Baghdad, and Al Basrah) as a Class A mandate entrusted to Britain. In the mandate system, a territory that had formerly been held by Germany or the Ottoman Empire was placed nominally under the supervision of the League of Nations, and the administration of the mandate was delegated to one of the victorious nations until the territory could govern itself. Mandates given Class A statuses were expected to achieve independence in a few years. While attending a conference in San Remo, Italy, in April 1920 the allied governments confirmed the creation of the British mandate in Iraq. By July, when the Iraqi Arabs learned of the decision, they began an armed uprising against the British, who were then still occupying Iraq.

The British were forced to spend huge amounts of money to suppress the revolt, and the government of Britain concluded that it would be advantageous to end its mandate in Mesopotamia. The British civil commissioner, the highest-ranking administrator in Iraq, quickly established a plan for a provisional government for the new state of Iraq, which would become a kingdom. A government directed by a council of Arab ministers under the supervision of a British high commissioner was established. Faisal was chosen to become the ruler of the new state, a decision confirmed by an election held in August 1921 in which he received 96 percent of the votes.

Faisal desperately needed to establish a local power base in

Iraq. He did this primarily by engaging the support of both Iraqi-born military officers who had served in the Ottoman army and Sunni Arab business and religious leaders in Baghdad, Al Basrah, and Mosul. Sunni Arab leaders and some Kurdish chieftains came to dictate the government and the army, whereas the Shi'a Arab chieftains and, to a lesser extent the Sunni Arab chieftains, came to control the parliament, enacting laws that were advantageous to themselves. Lower classes—poor peasants and Western-educated young men who depended on government jobs—had no input on affairs of the state.

Both the ruling elite (educated or privileged class) and the lower classes supported the pan-Arab movement, which sought to join all the Arab lands into one powerful state. The Pan-Arab movement was thought to be an effective means of uniting most of the diverse Iraqi population through their common Arab identity. The goals of Pan-Arabism were seen as being achievable through two means: the elite advocated diplomacy with British consent and the revolutionary and radically anti-British ideology of the lower classes.

Groups that didn't want to be part of the Arab movement, such as the Shi'as of the Euphrates River area and the Kurdish tribes of the north, challenged the integrity of the newly established state of Iraq. These groups acted in conjunction with Turkish armed forces hoping to reclaim the lands of the Mosul district for Turkey. The British were forced to maintain armies in Iraq, and aggression against the British mandate continued. King Faisal I requested that Iraq's mandate (authoritarian) system be converted into a treaty of alliance between the two nations. Britain chose not to end the mandate, although by June 1922, a 20-year treaty of alliance and protection between Britain and Iraq was signed. This treaty stated that the king would follow British advice on all matters affecting British interests and that British officials would serve in specific Iraqi government posts. In return, Britain agreed to provide military assistance and general aid to Iraq. The Iraqi national army was also created and would soon become an essential tool of domestic control for the ruling elite.

By the spring of 1924, a constituent assembly was convened. Laws establishing the permanent form of the government of Iraq were instituted. The king was given immense, but not absolute, power; he could dismiss parliament, call for new elections, and appoint the prime minister. In March 1925, elections for the first Iraqi parliament were held. That same year, an internationally owned oil company was granted permission to develop oil reserves in the Baghdad and Mosul regions. In 1927, Faisal I asked the British to support Iraq's application for admission to the League of Nations. Although the British refused at that time, by June 1930 a new treaty of alliance between Britain and Iraq included a recommendation by Britain that Iraq be admitted to the League of Nations as a free and independent state in 1932. The recommendation was made that year, and the British mandate system was formally terminated. In October 1932, Iraq joined the League of Nations as an independent sovereign (self-ruling) state. Faisal I died in 1933 and was succeeded by his son, Ghazi, a radical pan-Arab and anti-British figure.

**Iraq Becomes an Independent State**

In 1931, an agreement was signed by the Iraqi government and the Iraq Petroleum Company that further exploited the country's oil reserves. The Iraq Petroleum Company was an internationally owned organization composed of Royal-Dutch Shell, the Anglo-Persian Oil Company, French oil companies, and the Standard Oil companies of New York and New Jersey. The agreement gave the Iraq Petroleum Company the sole right to develop the oil fields of the Mosul region. In return, the company guaranteed to pay the Iraqi government annual royalties. The company had opened an oil pipeline from Mosul to Tripoli, Lebanon by 1934, and a second one to Haifa, in what is now Israel, was completed in 1936.

King Ghazi, Faisal's son (1933).

In 1936, Iraq, under King Ghazi, moved toward a pan-Arab alliance with the other nations of the Arab world. A treaty of nonaggression, reaffirming a fundamental Arab kinship, was signed with the king of Saudi Arabia the same year. Iraq experienced its first military coup d'état (overthrow) in 1936, when the army overthrew the pan-Arab Sunni government. The coup opened the door to future military involvement in Iraqi politics. The leaders of the coup included a Kurdish general and a Shi'a politician, and the moderate government they put in power was accepted by the king and remained in office until 1939. In April 1939, King Ghazi was killed in an automobile accident, leaving his three-year-old son, Faisal II, the titular king under a regency. Between 1936 and 1941, there were seven attempts by different factions to seize power. The last attempt was sponsored by the Germans in April 1941, due in part to their displeasure with Iraq's alliance with the British in 1939.

**Wars with Britain and Germany**

Adhering to its treaty of alliance with Britain, Iraq broke off diplomatic relations with Germany early in September 1939, at the start of World War II (1939–1945). For the initial months of the war, Iraq had a pro-British government under General Nuri as-Said as Prime Minister. Said was replaced in March 1940 by Rashid Ali al-Gailani, a radical nationalist, who immediately established a policy of noncooperation with the British. The British pressured the Iraqis to cooperate with them. This action brought about a military revolt on April 30, 1941, and a new pro-German government headed by Gailani was founded. The British, angered by this development, sent troops into Al Basrah. The military action by the British was declared a violation of the treaty between Britain and Iraq. Gailani mobilized the Iraqi army, and war between the two countries began in May. By the end of the month Iraq admitted defeat, and the terms of the armistice provided for the reestablishment of British control over Iraq. A pro-British government headed by Said was soon instituted. By 1942, Iraq had become a strategic supply center for British and United States forces operating in the Middle East and for the shipment of arms to the Union of Soviet Socialist Republics (USSR). On January 17, 1943, Iraq declared war on Germany, the first independent Islamic state to do so. At the same time, Iraq's continued assistance to the allied war effort enabled Arab leaders to make a stronger stand on behalf of a federation of Arab states. At the war's conclusion, the Arab League was formed when Iraq joined with other Arab nations in forming a regional association of sovereign states.

**Transjordan and Iraq**

Between 1945 and 1946, the Kurdish tribes that inhabited northeastern Iraq were in a state of unrest—supported, it was believed, by the USSR. Fearing that the Soviets would advance on the oil fields of Iraq, the British moved troops into the area. Tired of the meddling of foreigners, Prime Minister Said supported a new proposal in 1947 for a federated (united) Arab state. This time he suggested that Transjordan (present-day Jordan) and Iraq be united, and he began to negotiate with the king of Transjordan regarding the proposal. In April 1947, the two kingdoms, providing for mutual military and diplomatic aid, signed a treaty of kinship and alliance. With the establishment of Israel as a state in the same area of Palestine that the Arabs considered to be their homeland, the two Arab allies became enraged. In May 1948, the armies of Iraq and Transjordan invaded the new state. Throughout the rest of the year Iraqi armed forces continued to fight the Israelis, and the nation continued to work politically with the kingdom of Transjordan. In September, Iraq joined Abdullah ibn Hussein, king of Transjordan, in denouncing the establishment of an Arab government in Palestine as being "tantamount to recognizing the partition of Palestine" into Jewish and Arab states, which Iraq had vehemently opposed. The Arab forces that attacked Israel were defeated, and the government of Iraq prepared to negotiate an armistice, represented by Transjordan. On May 11, 1949, a cease-fire agreement between Israel and Transjordan was signed, but Iraqi units continued to fight Israelis in an Arab-occupied area in north central Palestine. Finally, Transjordanian troops replaced the Iraqi units in this area under the terms of the armistice agreement, signed on April 3, 1949.

### Founding of a Pan-Arab Government

The first parliamentary elections based on direct voting took place on January 17, 1953. These elections resulted in the founding of a pro-Western, pan-Arab government. King Faisal II formally assumed the throne on May 2, 1953, his 18th birthday. By this time Iraq had made significant progress in its political, social, and economic development. This was due in part to the wise management of the revenue received from the sale of its petroleum resources. There was a growing sense of nationalism in the young nation and a feeling that there should be less friendly relations with Western nations such as Britain. Iraq's leadership decided to follow Egypt's example: assert its independence and dedicate itself to support of a Pan-Arab state.

Still, interaction with the West continued in some regards. In February 1955, Iraq formed the Baghdad Pact, a treaty establishing mutual defense with Turkey. Plans were drawn to transform the alliance into a Middle Eastern defense system, and Iraq and Turkey urged the other Arab states, the United States, Britain, and Pakistan to stick to the terms of the pact. Britain joined the alliance in April; Pakistan signed on in September and Iran joined in November. The five nations established the Middle East Treaty Organization.

Iraq once again entered regional conflict in July 1956, when Jordan (Transjordan had been renamed) accused Israel of organizing an invasion army near Jerusalem, whereupon Iraq moved forces to the Jordanian border. Meanwhile, in response to Egypt's nationalization of the Suez Canal, which Britain and France had controlled, the Iraqi government expressed clear support of Egypt. In the Suez Crisis that followed, Egypt was invaded by Israel, Britain, and France in October 1956. Within a week, however, the United Nations, urged by both the USSR and the United States, demanded a cease-fire, forcing Britain, France, and Israel to withdraw from the lands they had captured. In early November, according to the terms of a mutual defense agreement, Iraqi and Syrian troops occupied positions in Jordan. In January 1957, Iraq endorsed the recently publicized Eisenhower Doctrine. The doctrine stated that the United States would supply military assistance to any Middle Eastern government whose stability was threatened by Communist aggression.

**Iraq Becomes a Republic**

In February 1958, Iraq and Jordan were united, following a conference between Faisal II and Hussein I, King of Jordan. The new union, which became known as the Arab Union of Jordan and Iraq, was established as a countermeasure to the United Arab Republic (UAR), a federation of Egypt and Syria formed in February of that year. The constitution of the newly formed federation was proclaimed simultaneously in Baghdad and Amman on March 19, and the document was approved by the Iraqi parliament on May 12. Later that month Nuri as-Said, former Prime Minister of Iraq, was named premier of the Arab Union. The UAR bitterly opposed the pro-Western Arab Union and issued repeated radio calls urging the people, police, and army of Iraq to overthrow their government.

A sudden coup occurred on July 14, 1958, led by Iraqi General Abdul Karim Kassem, and the country was proclaimed a republic. King Faisal II, the crown prince, and Premier Said were among those killed in the uprising. On July 15, the new government announced the establishment of close relations with the UAR and the disbanding of the Arab Union. General Kassem attempted to soothe and gain the confidence of the West by maintaining the flow of oil.

In March 1959, Iraq withdrew from the Baghdad Pact, which was then renamed the Central Treaty Organization; in June 1959, Iraq withdrew from the sterling bloc (a group of countries whose currencies are tied to the British pound sterling). The British protectorate over Kuwait ended in June

1960, and Iraq claimed the area, asserting that Kuwait had been part of the Iraqi state at the time of its formation. British forces entered Kuwait in July at the invitation of the Kuwaiti ruler, and the UN Security Council declined an Iraqi request to order their withdrawal. In domestic affairs, Iraqi government claimed in 1961 and 1962 that it had kept down Kurdish revolts in northern Iraq. The Kurdish unrest continued, however. The lengthy conflict was tentatively settled in early 1970, when the government agreed to form a Kurdish independent region and Kurdish ministers were added to the cabinet. This created an uneasy peace between the Kurds and the Iraqi government, but it did serve as a temporary solution.

On February 8, 1963, General Kassem was overthrown by a group of officers, most of them members of the Ba'th Party. He was assassinated the following day. Abdul Salam Arif became president and relations with the Western world improved. In April 1966, Arif was killed in a helicopter crash and was succeeded by his brother, General Abdul Rahman Arif.

**Conflicts with Israel**

Iraq entered the Arab-Israeli Six-Day War (1967) by sending troops and planes to the Jordan-Israeli border. Iraq's war with Israel began, and Iraq closed its oil pipeline supplying the Western nations because Iraqis believed that they sided with Israel. Diplomatic relations with the United States were stopped during the conflict. In July 1968, Ba'th Party officers overthrew General Arif's government. Major General Ahmed Hassan al-Bakr, a former prime minister, was appointed to head the newly established Revolutionary Command Council (RCC), the country's supreme executive, legislative, and judicial body.

In the years ahead, Iraq remained hostile toward the West and friendly with the USSR. The positions of individual Arab countries with regard to the legitimacy of the existence of Israel

General Abdul Karim Kassem, head of the new Iraqi republic in 1958. He was assassinated in 1963.

caused some friction between Iraq and its neighbors. In 1971, Iraq closed its border with Jordan and called for Jordan to be removed from the Arab League because of its efforts to crush the Palestinian opposition operating inside its borders. Iraq aided Syria with troops and supplies during the Arab-Israeli War of 1973. Iraq denounced the cease-fire that ended the 1973 conflict, called for continued aggression against Israel, and opposed the interim agreements negotiated by Egypt and Syria with Israel in 1974 and 1975.

Early in 1974, intense fighting began in northern Iraq between government forces and Kurdish nationalists, who rejected as inadequate a new Kurdish law granting independence to the Kurds based on the 1970 agreement. Led by Mustafa al-Barzani, the Kurds received support from Iran. After Iraq agreed in early 1975 to make major concessions to Iran in settling their border disputes, Iran halted aid to the Kurds, and the revolt was dealt a crushing blow. In July 1979, President Bakr was succeeded by General Saddam Hussein, a Sunni Muslim and fellow member of the Arab Ba'th Socialist Party. Hussein strengthened the country's internal security and eliminated all potential opponents. Within a short time after he came to power, Hussein's dictatorship had been firmly established.

In 1979, Iraq tried unsuccessfully to unite with Syria. Meanwhile, across the border in Iran, Islamic revolutionaries overthrew the country's secular (non-Islamic) government and established an Islamic republic. Tension between the Iraqi government and Iran's new Islamic regime increased during that year, when unrest among Iranian Kurds spilled over into Iraq. Increasing animosities between the Sunni-Shi'a religious groups increased the conflict.

**Iran-Iraq War**

In September 1980, Iraq declared null and void its 1975 agreement with Iran, which drew the border between the countries down the middle of the Shatt-al-Arab. Moreover, Iraq claimed authority over the entire river. The disagreement erupted into a full-scale war, the Iran-Iraq War. Iraq quickly conquered a large part of the Arab-populated province of Khuzestan (Khuzistan) in Iran and destroyed the Abadan oil refinery. In June 1981, a surprise air attack by Israel destroyed a nuclear reactor near Baghdad, with accusations by the Israelis that the reactor was intended to develop nuclear weapons for use against them. In early 1982, Iran launched a counteroffensive and by May had reclaimed much of the territory conquered by Iraq in 1980. Each side inflicted heavy damage on the other and on Persian Gulf traffic and shipping.

A cease-fire with Iran came into effect in August 1988, and the Iraqi government again moved to suppress the Kurdish revolt. In the late 1980s, the nation rebuilt its military machine, in part through bank credits and technology obtained from Western Europe and the United States.

### Oil Dispute with Kuwait and Other Military Crises

Iraq renewed a long-standing territorial dispute with Kuwait in the 1990s. Although Kuwait had been Iraq's ally during the war with Iran, Iraq claimed that overproduction of petroleum by Kuwait was injuring Iraq's economy by depressing the price of crude oil. Iraqi troops invaded Kuwait on August 2 and rapidly took over the country. The UN Security Council issued a series of resolutions that condemned the occupation, imposed a broad trade restriction on Iraq, and demanded that Iraq withdraw unconditionally by January 15, 1991. Iraq did not comply with the demands of the United Nations. A coalition of nations led by the United States began an intense bombing campaign that targeted military sites and strategic resources in Iraq and Kuwait in January 1991.

The resultant Persian Gulf War proved to be an international embarrassment for Iraq, which was forced out of Kuwait in about six weeks. Coalition forces invaded southern Iraq, and tens of thousands of Iraqis were killed. Many of the country's armored vehicles and artillery pieces were destroyed, and its nuclear and chemical weapons facilities were severely damaged. In April, Iraq agreed to UN terms for a permanent cease-fire; coalition troops withdrew from southern Iraq as a UN peacekeeping force moved in to police the Iraq-Kuwait border. Meanwhile, Hussein used his

Kurdish refugees after fleeing from Iraq to Isik Mountain, Turkey.

remaining military forces to keep down rebellions by Shi'as in the south and Kurds in the north. Hundreds of thousands of Kurdish refugees fled to Turkey and Iran, and US, British, and French troops landed inside Iraq's northern border to establish a Kurdish unit with refugee camps to protect another 600,000 Kurds from Iraqi government reprisals. In addition, international forces set up "no-fly zones" in both northern and southern Iraq to ensure the safety of the Kurdish and Shi'a populations.

The UN trade embargo remained in place even after the war. Strict demands were placed on Iraq to have these restrictions lifted, including destruction of its chemical and biological weapons, termination of nuclear weapons programs, and acceptance of international inspections to ensure that these conditions were met. Iraq claimed that its withdrawal from Kuwait was enough and resisted meeting the demands. Tensions with the United States continued to grow, and in June 1993 the United States launched a widely criticized cruise missile attack against Iraq in retaliation for a reported assassination plot against then US President George H.W. Bush.

Saddam Hussein signed a decree formally accepting Kuwait's political independence in 1994. The decree effectively ended Iraq's claim to Kuwait as a province of Iraq and was intended to curtail further tension in the region. Regarding affairs within the country, in 1994 Iraq continued its efforts to defeat internal resistance with an economic embargo of the Kurdish-populated north and a military campaign against Shi'a rebels in the southern marshlands. The Shi'as were quickly defeated, but the crisis in the Kurdish region was prolonged, not only between the Iraqis and the Kurds, but also among the Kurdish factions themselves. In the mid-1990s, clashes between the Patriotic Union of Kurdistan (PUK) and the Kurdistan Democratic Party (KDP) led to a state of civil war. Leaders of the KDP asked Hussein to intervene in the war, and he sent at least 30,000 troops into the Kurdish compound protected by international forces, capturing the PUK stronghold of Irbil. International forces decided to leave the area rather than intervene in the dispute between rival Kurdish factions.

The KDP was victorious and was quickly installed in power. The United States responded to Hussein's invasion with two missile strikes against southern Iraq. But the following month Iraq again helped KDP fighters, this time taking the PUK stronghold of Sulaymaniyah, and, by 1997, the KDP ruled most of northern Iraq. Relations had improved between the two Kurdish factions by September 1998, when the PUK and KDP reached an agreement that called for the founding of a joint regional government. Progress was initially slow, but the agreement eventually resulted in an end to armed conflict between the two groups.

Iraq faced a serious economic crisis in 1995 and 1996. Prices were high, food and medicine shortages were rampant, and the exchange rate for the dinar, Iraq's monetary unit, declined drastically. Sanctions against Iraq continued, although in April 1995 the UN Security Council voted unanimously to allow Iraq to sell limited amounts of oil to meet its urgent humanitarian needs. Initially rejecting the plan, Iraq finally accepted the conditions in 1996 and began to export oil at the end of that year. In 1998, the UN increased the amount of oil that Iraq was allowed to sell, but Iraq was unable to take full advantage of this increase because its production capabilities had weakened under the sanctions.

By May 1998, Saddam Hussein's interference with UN weapons inspectors nearly brought Iraq into another military crisis. The crisis was averted when UN Secretary General Kofi Annan negotiated an agreement that guaranteed Iraq's cooperation with the UN and avoided military strikes by the United States and its allies. Conflict arose again by December of that year, in response to reports that Iraq was continuing to block inspections. The United States and Britain launched a series of air strikes lasting for less than a week and damaged Iraqi military and industrial targets. In response to these attacks, Iraq declared that it would no

Scott Ritter of the United Nations waits to start his weapons inspection in Iraq on January 13, 1998. Iraq has continued to prevent these inspections.

longer comply with UN inspection teams, called for an end to the sanctions, and threatened to fire on aircraft patrolling the "no-fly zones." Through 2001, Iraq continued to challenge the patrols, and British and US planes struck Iraqi missile launch sites and other targets.

Kurds dressed in their traditional costumes dance just before a wedding.

# CHAPTER 4

# People and Culture

## Population

Iraq's population of 23.6 million includes about 75 percent Arab, 15 to 20 percent Kurdish, and a small percentage of minorities such as ethnic Assyrians, Turkomans, Armenians, Persians, Jews, and the Sabean and Yazidi religious sects. Iraqis tend to feel a common national sentiment, and most of them share a common language and religion, in spite of differing racial origins. Given the long history of Iraq, this is not surprising. Iraq has had many conquests and occupations by other peoples and cultures.

The overall population density of Iraq stands at 138 persons per square mile (53 per square km). But the population density varies greatly across the country, with the largest populations of this arid land concentrated near the river systems. About 76 percent of the population resides in urban areas, whereas in the rural sections of the country most people still live in tribal communities. Population growth rates, which averaged 3.2 percent per year in the early 1980s declined by the early 1990s as the birth rate fell during the Iran-Iraq War and the Persian Gulf War. By the late 1990s, however, the population growth rate had regained its former level. For 2001, the rate of population growth was 2.84 percent, the birth rate was 34.6 per 1,000 persons, and the death rate was 6.2 per 1,000 persons.

## The Kurdish People

In the mountainous north and northeast of Iraq and in and around the cities of Mosul, Erbil, Kirkuk, and Sulaymaniya live the Kurds, from the Persian *gurd,* or hero. The Kurds have a very strong sense of their own culture and traditions and are a proud but restless group of people who have faced persecution throughout the region. Traditionally, they have been semi-nomadic pastoral tribesman, who rear sheep and goats in the hills and valleys of Iraq's border country between Turkey and Iran. The Kurds speak their own language (which is a form of Persian, with two dialects) and have a keen sense of their own heritage. Kurdish people are found in an area that is divided among Turkey, Iran, Iraq, Syria, and Russia and is sometimes called "Kurdistan."

Although political maps of the region do not show it, where Iran, Iraq, and Turkey meet, the cultural landscape is not Iranian, Iraqi, or Turkish. This is the makeshift homeland of the Kurds, a nation of about 25 million (their numbers are uncertain). More Kurds live in Turkey than in any other country (as many as 12 million); possibly as many as seven million live in Iran; about half that number live in Iraq; and smaller numbers live in Syria, Armenia, and Azerbaijan.

For over 3,000 years, the Kurds have occupied this isolated, mountainous frontier. The Kurds are a nation, but they have no state, and they do not receive the international attention that other stateless nations have received (such as the Palestinians). They have been repressed by Turks and Iraqis. Certainly their landlocked location has something to do with their dismal situation, since its remoteness and the barriers created by the ruling powers hinder their access to getting international aid.

The Kurdish people have long dreamed of a day when their splintered homeland will be a nation-state. The city of Diyarbakir, now in southeastern Turkey, would most likely become the capital, although the largest urban concentration of Kurds today is in the shantytowns of Istanbul, Turkey, where as many as 3 million Kurds have migrated. Without a territory of their own, without political power, and without the ability to shed light internationally on their problems, the Kurds may wait forever for the realization of their dream.

**Marsh Arabs**

A tribal group of dark-skinned people known as the Marsh Arabs resides in the vast area of marsh and lagoons near the area in southern Iraq, where the Tigris and Euphrates rivers join. These marsh dwellers most likely came from the groups of ex-slaves and outcasts who first sought refuge and anonymity during a 19th-century rebellion in the area. The way of life for the Marsh Arabs has changed little through the centuries, and only in the last few decades have these areas even been mapped and explored by foreigners. The Marsh Arabs remain largely a mysterious people even to their fellow countrymen.

The Marsh Arabs have in recent years begun to follow the events of the modern world. Iranian influence has grown in this region of Shi'a Muslims, and a quiet revolt against the regime of Saddam Hussein has been taking place here since the end of the Persian Gulf War in 1991. Having noticed the silent uprising of the marsh dwellers, Hussein had many of the marshes drained at the end of the war to flush out the rebels who were taking cover in the marshes. Many fled to Iran.

Marsh Arabs hunt for food.

## Bedouins

The Bedouins are nomadic tribes who migrate seasonally between western and southern Iraq and between Iraq's neighboring countries of Kuwait and Saudi Arabia. Winters are often spent in the southern well-vegetated areas, whereas summers are spent near the oases on the edge of Iraq's western desert area. Camel breeding has been the focus of Bedouin livelihood, for camels provide a source of meat, hide, hair, and transportation. The

People and Culture 55

This Bedouin woman in Jordan wears traditional dress.

Bedouins live a modest and harsh life, and this has pressured the tribes-people to abandon their traditional existence for village life, where farming can be practiced. As a result, the number of Bedouins who live as nomads has declined, although Bedouins remain a common sight throughout the western areas of Iraq.

## Other Ethnic Groups

Modern Iraq is home to several other ethnic minorities. Found in the Kurdish mountains of northern Iraq, the Assyrians speak their own language (Syriac) and are followers of the Christian Nestorian Church. Although they were unhappy with being incorporated into the newly formed Republic of Iraq in 1933, they have integrated well into Iraqi society. Today, the Assyrians are largely accepted as equal members of society, as are the Armenians—Christian refugees from the Ottoman Turks. A large Armenian population is found in Baghdad, and they have their own language, religion, and trading community.

Near the northern city of Kirkuk lives a large community of Turkomans. They are racially pure descendants of the Mongol invaders from central Asia who were led by the Turkish leader Tamerlane to conquer Iraq nearly 700 years ago. In addition, a small number of people of Persian (Iranian) descent settled in and around the Shi'a holy cities of Najaf, Kerbala, Samarra, and Kadhimain. During the 1980s war with Iran, many fled back to their country of origin.

In northwest Iraq near Jebel Sinjar live the Yazidis (from the Persian *Yazdan*, meaning God). Interesting groups of religious people, the Yazidis speak a Kurdish dialect, use Arabic in their worship, and regard both the Koran and the Bible as holy books. The Yazidis also practice snake charming and snake swallowing.

The Sabeans, sometimes called the Christians of St. John because of their claim to be followers of John the Baptist, are another minority group found in Iraq. Their claim is inaccurate, however, since their beliefs are closely tied to those of the Muslim faith and they are mentioned in the Koran. Originally dwelling near the riverbanks in southern Iraq, they are master boat builders. This is not surprising because their religion requires that they live beside running water.

## Religion

Muslims (followers of Islam) make up about 97 percent of Iraq's population, with about 55 to 65 percent of the Muslims adhering to the Shi'a branch and the rest adhering to the

Iraqi women walk near a sacred mosque in Kerbala, Iraq.

Sunni branch of Islam. The Shi'a are found primarily in central and southern Iraq, and the Sunnis principally reside in the north. Most Kurds are Sunni, as are most of the country's political leaders.

The word *Islam* means "submission to the will of God" (Allah in Arabic). Islam, like Christianity and Judaism, is monotheistic (worships one God). The faith of Islam is summed up in the Shahada: "There is no God but God, and Muhammad is his prophet." Saying the Shahada once, aloud, with understanding and belief is necessary to become a Muslim. The Koran is the holy book that contains the inspired words of Muhammad given by God through the Angel Gabriel. The Muslims honor this book and believe it contains what they need for salvation. Also, the Muslims believe that the Christian gospels, the first five books of the Old Testament, and the Psalms are the inspired words of Muhammad. They do not believe that the texts we have today are as God gave them. Muslims do recognize Adam, Noah, Abraham, Moses, and Jesus as great prophets, but for them Muhammad is the true, last, and greatest prophet.

The Hadith contains traditions based on sayings and deeds of Muhammad. Along with the Koran, the Hadith provides the basis for the code of behavior that governs Muslims. That code is called Sunna. The two main groups of Muslims, the Sunnis and the Shi'as, each have different Sunna.

Muslims are required to pray five times a day while facing Mecca. Next to the mosques—buildings set aside for group prayer—are minarets, or towers. From these towers, a person calls the faithful to prayer at the required times. Prayers offered on Fridays are considered to be most pleasing to Allah. Friday also is the day when Muslims gather at their mosques to pray and hear a sermon. The preachers, called *imams*, are teachers, not priests with religious authority to stand between the worshiper and God.

Fasting is required during the holy month of Ramadan. No food, drink, tobacco, or other worldly pleasure may be taken from dawn until sunset. Ramadan falls at different times in different years because it is determined by a calendar based on the moon. When Ramadan falls during the long, hot days of summer, fasting is not easy. Exceptions are made for the sick, the weak, soldiers, and travelers. Giving to the poor and the mosques is also a requirement. Even with state-provided welfare system, generosity to others is highly valued.

Pilgrimage to Mecca is expected of every Muslim who is able (physically or financially) to go. Only Muslims are allowed in the holy cities of Mecca and Medina in Saudi Arabia. Pilgrims who make the journey have an elevated status on the community upon their return.

As with other religions, adherence to these practices varies. For a time, Muslims who had been influenced by other cultures took a more relaxed view of these duties, and so did the poorer working classes. In recent years, there has been a renewed interest in following these practices more strictly throughout the Arab world. Iraq took a liberal view of these requirements at the beginning of the revolutionary movement, but now the government seems to be taking a more orthodox stand. The phrase *Allahu Akbar* (God is Great) in green Arabic script—*Allahu* to the right of the middle star and *Akbar* to the left of the middle star—was added to Iraq's flag in January 1991 during the Persian Gulf War crisis.

Small communities of Jews call Iraq home. These people trace their ancestry back to the Babylonian exile of the Jewish people that occurred from 586 to 516 B.C. The Iraqi Jews have declined in number in recent years.

Christians are the largest religious minority in Iraq today. They are descendants of the people who did not convert to Islam during the Arab conquest of the sixth and seventh centuries. The Assyrians, for example, are descendants of the Mesopotamians, and speak Aramaic. Some have dispersed throughout the United States and Canada.

Members of Iraqi National Assembly await the morning session of Parliament in Baghdad, January 9, 1999.

# 5

# Government

Emerging as an independent nation after the fall of the Turkish Ottoman Empire in 1918, Iraq was a monarchy (ruled by one person) from 1921 to 1958, when military officers overthrew the monarchy in a bloody coup d'état (revolution) and set up as a republic. Since 1968, the government has been a dictatorship dominated by a single political party, the Ba'th Party (the only officially recognized party), and the people have little or no say in their government. The Ba'th rule Iraq through a nine-member Revolutionary Command Council (RCC), which enacts legislation by formal decree. The RCC's president, who also serves as the chief of state and supreme commander of the armed forces, is elected by a two-thirds majority of the RCC. Saddam Hussein has held this position since 1978. The council also selects a cabinet that has some administrative and legislative responsibilities. There are occasional elections to the legislature, and the president was once "confirmed" in 1995 in a public referendum, but none of these seemingly democratic procedures was truly democratic. In reality, the people do not elect their rulers, since only Ba'th Party–authorized candidates can run for election. The Iraqi regime simply does not tolerate resistance, and therefore opposition parties either operate illegally as exiles from neighboring countries or in areas of northern Iraq outside the control of the regime.

The Kurdish Democratic Party led by Masoud Baranzi and the Patriotic Union of Kurdistan led by Jalal Talabani are opposition parties, each of which control portions of northern Iraq. Both groups allow multiple political parties to operate in their territories. Each has held elections in the past year that most international observers have deemed to be generally fair. The two most important Shi'a opposition parties are the Da'wa Islamic Party and the Supreme Assembly of the Islamic Revolution in Iraq. These parties are illegal outside the Kurdish autonomous (independent) region.

From 1968 to 1978, under President Ahmed Hassan al-Bakr, the Ba'th Party ruled the country with an iron fist. Ironically, within the party there was a surprising degree of democracy, since in many cases people rose to positions of power from below in a semidemocratic way. Top-level discussions were conducted in a fairly free fashion, and often decisions were made through a discussion of party leaders and after taking a vote. In 1979, this semidemocracy ended after Saddam Hussein came to power, replacing al-Bakr as president. As is widely believed, Hussein forced his predecessor to resign. Immediately, Hussein had 55 senior party activists and army officers executed for treason, despite the fact that there was no real evidence of their having been guilty of treason. The most likely reason for their elimination was either opposition to Hussein's replacing al-Bakr or a dispute over the way in which Hussein would be

Sample ballot from Iraqi elections.

elected. Throughout Hussein's rule, more executions for disloyalty have followed, sending an unmistakable message that no one is to question his decisions.

Iraq is governed under a provisional (temporary) constitution that was adopted in 1969 and amended through the years. The constitution defines Iraq as "a sovereign people's democratic republic," dedicated to pan-Arabism—the ultimate realization of a single Arab state—and to the establishment of a socialist system. The document declares Islam as the state religion, but guarantees freedom of religion. It defines the Iraqi people as comprising two principal nationalities, Arab and Kurdish. A 1974 amendment grants autonomy for the Kurds in areas where they constitute a majority, but states that Iraq is to remain united and undivided. The state is given the important central role in "planning, directing, and guiding" the economy, and national resources are defined as "the property of the people." The constitution states that the RCC selects the

Iraqi President Saddam Hussein praises his military for their achievements in a televised speech commemorating Army Day and the 77th anniversary of the Iraqi army on January 6, 1998.

president, who is the official head of state. The president appoints all government, civil service, and military personnel and approves the budget. Most of the president's power comes from his role as chairman of the RCC. And there is no fixed term for president.

Iraq had no national legislature from 1958 until 1980,

when a National Assembly was established. This assembly consists of 250 members, 220 of whom are elected by popular vote and 30 are appointed by the president to represent the three northern provinces. Each member serves a four-year term. A candidate for the National Assembly must demonstrate "loyalty to the principles of the Ba'th (1968) Revolution"—meaning, the goals of the Ba'th Party—to an electoral commission. The main task of the legislature is to approve or reject legislation proposed by the RCC. Except in a few minor cases, however, the National Assembly is a mere rubber stamp. Draft laws suggested by the National Assembly must be approved by the RCC before they become laws, and the RCC can enact ordinances that have the force of law. Nevertheless, the National Assembly creates the appearance of democratic procedures.

Principal government figures as of December 2001 are Saddam Hussein (President, RCC Chairman, Prime Minister, Ba'th Party Regional Command Secretary General) and Taha Yasin Ramadan and Taha Muhyi al-Din Ma'ruf (vice presidents).

## Court System

Iraq's judicial system is based on the French model of judicial systems introduced during Ottoman rule. It has three types of lower courts—civil, religious, and special. Special courts try very broadly defined national security cases. There are five types of civil courts. The first are courts of justices of the peace, whose jurisdiction is limited to minor cases and which are located in the major cities of Baghdad, Al Basrah, Mosul, and Kirkuk. Next are courts of the sole judge, which hear more serious cases and are located in all provincial capitals and subdivisions. Courts of first instance hear major civil cases and are located in all the provincial capitals. Appellate courts are located in the same four major cities as are the courts of justices of the peace.

For criminal cases, a magistrate decides if the matter is a major offense or a lesser offense, and then investigates a complaint of a crime. Penal courts try the lesser offenses; such courts usually consist of one judge. Major offenses are tried in a great court, which usually consists of three judges. Criminal courts are located wherever there are civil courts.

The religious courts were abolished after the revolution of 1958, but were reinstated by the Ba'th regime in the 1980s. Religious courts are located wherever there are civil courts. They try cases of personal status, such as divorce cases and disputes involving *waqf*, which are gifts of land or property made by a Muslim and intended for religious, educational, or charitable use. There are Sunni, Shi'a, and Christian religious courts.

An appellate court system and Iraq's highest court, the Court of Cassation (court of last recourse) complete the judicial system. The Court of Cassation has 12 to 15 judges, who must have previously served as judge for at least 15 years before being appointed to this court. In the Court of Cassation, at least three judges must hear a case. In cases in which the offense is punishable by death, five judges are required. This court has several departments: general, civil, criminal, administrative, and personal status. Sometimes these departments may try high officials or judges. A minister of justice has jurisdiction over all the courts and nominates all judges, who are then appointed by the president.

## Provincial Government and Organizations

Iraq is divided into 18 provinces, three of which are designated officially as a Kurdish autonomous (independent) region. Outside the Kurdish region, a governor appointed by the national government administers each province. Councils headed by mayors run cities and towns. Established in 1970, the Kurdish autonomous region has an elected 50-member legislature. This region came under UN and coalition protection after the Persian Gulf War to prevent Hussein from taking military action against rebellious Kurds.

Internal bickering among Kurdish groups rendered the government largely ineffective. In 1998, two rival Kurdish parties signed an agreement, brokered by the United States, which provided for a transitional power-sharing arrangement. This agreement has not been put into effect though, and each of the two parties governs its own territory.

Iraq is a charter member of the United Nations and a founding member of the Arab League, whose energies in years past have been devoted to political, economic, and propaganda warfare against Israel. Recently, the Arab League has engaged in more constructive pursuits such as organizing peacekeeping forces among its members, sponsoring scholarship and training programs for Palestinian Arab refugees and engaging in some social and cultural activities. Iraq is also a founding member of the Organization of the Islamic Conference, which promotes solidarity among nations where Islam is an important religion, and the Organization of Petroleum Exporting Countries (OPEC).

An Iraqi money trader in January 1999 demonstrated how many Iraqi dinars (Iraqi monetary units) were the equivalent of US $20. The economic sanctions imposed after the Persian Gulf War so devalued the currency that it took a fistful of dinars just to buy groceries.

# CHAPTER 6

# Economy

The economy of Iraq, which was mostly traditional and nonprogressive until 1914, has changed remarkably in this century in terms of its scale and quality of agriculture, development of oil reserves, trade and industry, and transportation methods. The modern Iraqi economy has been largely based on petroleum. Most of the few large manufacturing industries have to do with oil.

Beginning in 1980, the Iraqi economy was unfavorably affected by four major factors: the war with Iran during the 1980s, an international oil surplus in the 1980s and 1990s, the economic sanctions imposed by the UN (after the invasion of Kuwait in 1990), and the Persian Gulf War in 1991. The combined effect of all these factors was the destruction of Iraq's basic infrastructure (roads, bridges, power grids, and the like) and the country's financial bankruptcy.

Studies done at the end of the 20th century revealed that Iraq's real gross domestic product (GDP)—that is, its GDP adjusted for inflation—fell by 75 percent from 1991 to 1999. Iraq has enormous economic potential. With its vast oil resources, population, and water supply, it could successfully develop many industries. Yet the dictatorship of Saddam Hussein has left the country in serious debt and its people deprived of basic foods and medicines.

Petroleum is the most important natural resource of Iraq. Iraq has an estimated 10 percent of the world's supply of proven petroleum reserves. The oil fields are located in two main regions: in the southeast, just inland from the Persian Gulf, near Ar Rumaila, and in the north central part of the country, near Mosul and Kirkuk. Minor deposits of various other minerals are found throughout the country, such as ores of iron, gold, lead, copper, silver, platinum, and zinc. Phosphates, sulfur, salt, and gypsum are fairly abundant, and seams of brown coal are numerous.

Heavy dependence on oil exports and a focus on central planning characterize the economy of Iraq today. Before the outbreak of the war with Iran in September 1980, Iraq's economic prospects were bright. Oil production had reached a level of 3.5 million barrels per day, and oil revenues were $21 billion in 1979 and $27 billion in 1980. Just before the fighting began, Iraq had collected an estimated $35 billion in foreign exchange reserves.

The effects of the war with Iran were ruinous. Iraq's foreign exchange reserves were depleted, its economy crumbled, and the country was burdened with a foreign debt of more than $40 billion. At the war's end, oil exports gradually increased with the construction of new pipelines and the repair and restoration of damaged facilities.

But Iraq's invasion of Kuwait in August 1990, subsequent international sanctions, and damage from military action by an international coalition (alliance) beginning in January

1991 drastically reduced economic activity. Hussein's policies of diverting income to key supporters of the regime while supporting a large military and internal security force further impaired finances, leaving the average Iraqi citizen facing desperate hardships. Since December 1996, the implementation of a UN oil-for-food program has given some relief from the sanctions and improved conditions for the average Iraqi citizen. However, this program was not going to solve the fundamental problems of a devastated economy and of a population impoverished by two successive wars and years of severe economic sanctions. Beginning in 1999, Iraq was authorized to export unlimited quantities of oil to finance humanitarian needs including food, medicine, and infrastructure repair parts. Oil exports fluctuate as the regime alternately starts and stops exports, but, in general, oil exports have now reached three-quarters of their pre-Gulf War levels, although per capita (per each individual) output and living standards remain well below pre-Gulf War levels.

To gain the support of other nations, Iraq must promise profitable post-sanction oil contracts to potential allies. Indications are that Russia, China, and France will be the main beneficiaries of these promises and will therefore support either softening or lifting of the sanctions. However, these control measures are not likely to be lifted as long as Saddam Hussein remains in power. Meanwhile, the Iraqi government is expected to focus on getting around the sanctions, primarily through oil smuggling. Estimates place current revenues from illegally sold oil in the million to billion-dollar range.

Iraq's real GDP was estimated to be in the 1990s about what it was in the 1940s before the oil boom and the modernization of the country. Per capita income (earnings of each person) and the people's food intake plunged from the levels of relatively better-off Third World countries (underdeveloped nations) to those of

Iraqi man carries his monthly food ration, which is not enough. He will need to buy more food to feed his family.

the desperately poor Fourth World states, such as Haiti and Rwanda, the Democratic Republic of the Congo, and Somalia in Africa. Fourth World countries are characterized by low per capita income and few valuable natural resources. Since the end of the Persian Gulf War, all aspects of Iraq's economy have been devastated. Its valuable assets, as well as its basic social and economic infrastructure, have been wasted or permanently destroyed. The well educated of Iraq have fled, and even the value of its national currency, the dinar, has continued to decline, driving prices upward. The government has foolishly continued to finance its spending commitments by printing money, thus guaranteeing that price increases would continue.

For Iraq, the early 1970s had been a time of important development for the Iraqi economy and the government's role in it. By 1972, the government had nationalized the Iraq Petroleum Company (IPC), which had been owned by foreign oil companies. The nationalization combined with the steep rise in the price of crude oil that the Organization of Petroleum Exporting Countries (OPEC) engineered in 1973, had the effect of raising Iraq's oil revenues more than eightfold—from $1 billion in 1972 to $8.2 billion in 1975. These substantial increases in revenues solidified the government's role in the economy, making the government mainly responsible for transferring wealth from the petroleum industry to the rest of the economy. Through this central role, the government acquired the power to distribute economic resources to various sectors (divisions) of the economy and among different social classes and groups. In the 1970s, the Iraqi government was the primary determiner of employment, income distribution, and development, both of economic sectors and of geographic regions. It developed extensive plans for the economy and exercised heavy control over agriculture, foreign trade, communication networks, banking services, public utilities, and industrial production, leaving only small-scale industry, shops, farms, and some services to private businesses.

### Exports and Imports

Before the UN-imposed trade restrictions on Iraq after Iraq's invasion of Kuwait in 1990, average annual earnings for exported products were estimated at $10.4 billion and for imports at $6.6 billion. Petroleum sales accounted for almost all the export earnings; other exports were dates, raw wool, and hides and skins. Primary imports were machinery, transportation equipment, foodstuffs, and pharmaceuticals. Iraq's main trade partners were Brazil, Turkey, Japan, Germany, France, Italy, the United Kingdom, and the United States. After the Gulf War, Iraq's trading partners have mainly included only Russia, China, France, and Egypt.

Predominantly an agricultural country, Iraq has about 12 percent of its land under cultivation, with most farmland being in and around the Tigris and Euphrates rivers. Agricultural production in 2000 included 384,000 metric tons of wheat, 226,000 metric tons of barley, and 130,000 metric tons of rice. Before the UN sanctions, exports of dates from Iraq accounted for a major share of world trade in dates. Other fruits produced include apples, figs, grapes, olives, oranges, pears, and pomegranates.

The raising of livestock is an important occupation for Iraq's nomadic and seminomadic tribes, with about 10 percent of Iraq's land area being suitable for grazing. In 2000, the livestock population included 1.1 million cattle, 6.1 million sheep, 1.35 million goats, and 19 million poultry. In addition, the world-famous Arabian horse is extensively bred, although demand for the breed has declined in recent years. In addition, Iraq has a small fishing industry. In 1997, 34,702 metric tons of fish were caught, mostly freshwater species.

### Industry

Manufacturing is not a well-developed industry in Iraq. Petroleum and natural gas products are produced, but manufactures are largely limited to goods such as processed foods and beverages, textiles and clothing, metal products, furniture, footwear, cigarettes, and construction materials. Baghdad is the leading manufacturing center of Iraq. Important sectors of the service industry include government social services such as health and education, financial services, and personal services.

Man works at al-Rumaila oilfield in the port of Basra in south Iraq.

**Money**

Iraq's official monetary unit is the Iraqi dinar (0.31 dinars equal US $1; fixed rate). The Central Bank of Iraq, which also controls the banking system and foreign exchange transactions, issues currency. In addition to the Central Bank, the system consists of the Rafidain Bank, which handles government accounts, including oil revenues, and five specialized banks: the Agricultural, Cooperative, Industrial, Mortgage, and Real Estate banks.

Iraqis reading newspapers on March 1, 2002. There is concern over Britain's accusations that Iraq has accumulated weapons of mass destruction that pose a threat to the world.

# CHAPTER 7

# Living in Iraq Today

Although Iraq has great potential because of its water and oil resources and relatively well-educated and skilled population, the repressive leadership of Saddam Hussein has had a stifling effect on the Iraqi people, who today endure great hardships.

Most of the ruling elite of Iraq are of the Sunni Arab population. There are very few Shi'a Arabs in the middle and upper classes of society, and poverty is widespread among them. The Kurds do not enjoy even the limited representation in the government that the Shi'a do. Since 1961, the northern Kurdish areas of Iraq have been involved in ongoing revolts against the government. Other than religion, Iraqi society faces yet another divide: urban versus rural populations. Although the pace of urbanization in the country has increased in recent years, many Iraqis, especially those in rural areas, maintain extended family relations and tribal connections. These relationships have been further strengthened by the recent economic hardships, because entire families must work together to earn even a meager living.

## Health Standards and Medicine

Iraq's health standards have declined because of the many endemic diseases (diseases native to the country), poor sanitary conditions, and lack of medicines. Basically, the sanctions against Iraq have resulted in the decline of health standards ever since the Persian Gulf War. In 2001, life expectancy at birth was 58 for males and 60 for females, both of which fall about 20 years short of the same figures for the United States. The infant death rate was approximately 92 per 1,000 live births. Iraq has only one physician for every 2,091 people and about one hospital bed for every 690 people. Most medical facilities, controlled by the central government, do not have enough supplies and need repairs and modernization.

The Iraqi government has tried to improve health standards. In September 1985, Iraq began the first stage of a national campaign to vaccinate Iraqi school children against tetanus, diphtheria, whooping cough, poliomyelitis, tuberculosis, measles, and German measles. Improvements in flood control, water supply, and sewage systems have gone a long way toward making the dreadful epidemics (in which thousands of people died) a thing of the past. In addition, government health education campaigns have improved people's understanding of the importance of hygiene. As with other sectors of society, the campaign to improve the health of Iraqi citizens has been affected by the UN sanctions, and some ground has been lost.

## Social Services and Education

There was a massive expansion in social services in Iraq during the 1980s. Many housing developments were completed, schools were opened in every village, and hospitals

Living in Iraq Today 79

School children share textbooks and desks at the Al-Wahda primary school, September 4, 2000.

and clinics were set up throughout the country. A major health campaign with the slogan *Health for All by the Year 2000* was launched. Since 1968, the numbers of doctors and dentists in Iraq has doubled, and the number of pharmacists has increased greatly. A lingering problem is a lack of qualified nurses, largely because of the very low status of nursing in Iraq. Nursing was always considered a dirty and humiliating job. To remedy this, the government required in 1985 that all females who left school or graduated from college must complete one year of nursing service before beginning other employment.

One of the government's most successful development programs is education. Almost every village in Iraq has at least an elementary school. And every town has a secondary school, although rural schools are often built of mud brick and have tin roofs. Education is provided free by the state. Six years of primary education are compulsory; still, many children do not attend, since they must work to support the family. The literacy rate averages about 74 percent among those 15 and older.

The origins of the Iraqi educational system are found in the traditional Koranic schools—religious institutions attached to every mosque (building for prayer). Throughout the centuries, these schools have provided a basic education in which scriptures are memorized and the interpretation of the religious instructors (*mullahs*) is never questioned. Even in modern Iraq, most students in schools and colleges concentrate on memorizing facts. They are taught to believe that the teacher is always right. A high value is placed on education, and a degree or diploma is very much a passport to a good job and a good salary.

Today, Iraq has eight universities (four in Baghdad and the others in Basra, Mosul, Tikrit, and Erbil) and 22 institutes of technology, agriculture, and administration. Education is given a high priority by the Iraqi government. The government recognizes that a literate and skilled population is essential if Iraq is to continue its social progress and its industrial and technological advancement.

## Road and Transportation Systems

Neither road nor railroad system was ever well developed in Iraq. The country has railroad connections through Syria with Turkey and Europe, with the Iraqi state railway system consisting of about 1,515 miles (2,438 km) of track. Nevertheless, the present system is poorly planned and is seldom used for passenger traffic. When the railroad system was constructed, differing widths were used for different sections of the railroad system. Therefore, today the national railroad system is not always compatible from one area to another. Plans are in development to add some new 2,000 miles (3,219 km) of tracks that would connect Iraq's major cities.

Iraq has about 29,453 miles (47,402 km) of roads. In 1997, the country had 52 motor vehicles in use for every 1,000 people; the rate for passenger cars was 36.3 per 1,000. Even before the devastation of the recent wars, the highways were strained by Iraq's growth. In a virtually landlocked country such as Iraq, ample roadways are very important. There are plans for upgrading and expanding the national highway systems, but this is not likely to occur until Iraq is released from the UN sanctions.

International airports serve Baghdad and al Basrah, although during times of war or other national crisis the government usually imposes a ban on international travel. Domestic air service exists between Baghdad, Mosul, and Basra. Basra is also important to Iraq's water transportation because it, and Umm Qasr, are the main ports for ocean-going vessels. River steamers are able to navigate the Tigris from al Basrah to Baghdad.

During the Persian Gulf War, bombing by United States-led coalition air forces destroyed many transport facilities, such as bridges, ports, and airports. Some estimates suggest that the bombing destroyed more than 80 bridges. Iraq was able to rebuild some bridges and other facilities in the years after the war. The bombing campaign combined with the UN control measures have crippled Iraq's plans to expand all areas of national transportation.

In Baghdad on January 17, 2002, Iraqi Foreign Minister Naji Sabree al-Hadithi (center) walks on a portrait of US President George W. Bush. That day, President Saddam Hussein declared that his country is ready for any possible attack by the United States in response to Bush's warning to allow UN arms inspectors into Iraq.

# Iraq Looks Ahead

The future of Iraq looks bleak. The country has largely been isolated politically and economically since the end of the Persian Gulf War and the imposition of UN sanctions. And the control measures are unlikely to be lifted as long as Saddam Hussein is in power and continues his oppressive dictatorship.

The United States Secretary of State has placed Iraq (along with Iran, Syria, Libya, Cuba, North Korea, and Sudan) on a list designating it as a "state that sponsors terrorism." This designation took on greater significance after the September 11, 2001 attacks on the United States. The designation—and the imposition of sanctions—is a mechanism for isolating nations that use terrorism as a means of political expression. United States policy seeks to pressure and isolate state sponsors of terrorism so that they will renounce the use of terrorism, end support to terrorists, and bring terrorists to justice for past crimes. The declaration by the State Department asserted,

> The United States is committed to holding terrorists and those who harbor them accountable for past attacks, regardless of when the attacks occurred. The US government has a long memory and will not simply expunge a terrorist record because time has passed. The states that choose to harbor terrorists are like accomplices who provide shelter for criminals. They will be held accountable for their guests' actions.

According to the State Department, Iraq planned and sponsored international terrorism in 2000 and 2001. Although Baghdad focused on "anti-dissident" activity overseas, the regime continued to support various terrorist groups. (Dissidents are people or groups who disagree with an established political or religious system.) Iraq has not attempted an anti-Western terrorist attack since its failed plot to assassinate former President George H.W. Bush in Kuwait in 1993.

Czech police continued to provide protection in the Prague office of the US Government–funded Radio Free Europe/Radio Liberty (RFE/RL), which produces Radio Free Iraq programs and employs expatriate journalists (writers who have left or renounced their home country). The police presence was increased in 1999 after reports that Iraqi Intelligence Service (IIS) might retaliate against RFE/RL for broadcasts critical of the Iraqi regime.

To intimidate or silence Iraqi opponents of the regime living overseas, the IIS reportedly opened several new stations in foreign capitals during 2000. Various opposition groups joined in warning Iraqi dissidents abroad against newly established "expatriate associations," which, they asserted, are IIS front organizations. Opposition leaders in London contended that the IIS had dispatched women agents to infiltrate their ranks and were targeting dissidents for assassination. In Germany, an Iraqi opposition figure denounced the IIS for murdering his son, who had recently left Iraq to join him abroad. Dr. Ayad Allawi, Secretary General of the Iraqi National Accord, an opposition group, stated that relatives of dissidents living abroad are often arrested and jailed to intimidate activists overseas.

In northern Iraq, Iraqi agents reportedly killed a locally well-known religious personality who declined to echo the regime line. The regional security director in As Sulaymaniyah stated that Iraqi operatives were responsible for the car bomb explosion that injured a score of bystanders. Officials of the Iraqi Communist Party asserted that an attack on a provincial party headquarters had been thwarted when party security officers shot and wounded a terrorist employed by the IIS.

Baghdad continued to denounce and de-legitimize UN personnel working in Iraq, particularly UN teams clearing land mines, in the wake of the killing in 1999 of an expatriate UN de-mining worker in northern Iraq under circumstances suggesting regime involvement. An Iraqi who opened fire at the UN Food and Agriculture Organization (FAO) office in Baghdad, killing two persons and wounding six, was permitted to hold a heavily publicized press conference at which he contended that his action had been motivated by the harshness of UN sanctions, which the regime regularly attacks.

The Iraq regime rejected a request from Riyadh for the extradition of two Saudis who had hijacked a Saudi Arabian Airlines flight to Baghdad, but who did return promptly the passengers and the aircraft. Disregarding its obligations under international law, the regime granted political asylum to the hijackers and gave them ample opportunity to vent their criticisms of alleged abuses by the Saudi Arabian government in the Iraqi-government–controlled and international media. This echoed an Iraqi propaganda theme. While the origins of the FAO attack and the hijacking were unclear, the Iraqi regime readily used these terrorist acts to further its policy objectives.

Several expatriate terrorist groups continue to maintain offices in Baghdad, including the Arab Liberation Front, the inactive 15 May Organization, the Palestinian Liberation Front (PLF), and the Abu Nidal Organization (ANO). PLF leader Abu Abbas appeared on state-controlled television in the fall of 2000 to praise Iraq's leadership in rallying Arab opposition to Israeli violence against Palestinians. Many charges have been presented against Iraq by the United States and government officials have repeatedly mentioned Iraq as a potential future target in the war on terrorism.

Currently, Iraq's main resistance group asserts that internal opposition to Saddam Hussein is so strong that it would not take much effort to remove him from power. The plan is being circulated by the Iraqi National Congress (INC), a London-based confederation of Iraqi opposition groups that enjoys considerable support from the United States Congress, but is seen as largely ineffectual by many in President George W. Bush's administration. The opposition group believes that an Afghanistan-style bombing campaign and the insertion of United States Special Forces would cause mass defections and crumble Saddam's regime.

"What happened in Afghanistan is basically what we want to do in Iraq," says Ahmad Chalabi, leader of the INC. Chalabi also states that Iranians are prepared to help the United States remove Saddam from power. Specifically, the INC states that Iran will provide transit, staging, and logistical support for Iraqi rebel troops if the United States fully supports to the operation's success. Some in the administration believe that the United States should seek help from Iraqi neighbors other than Iran—such as Turkey, Jordan, and Saudi Arabia. Whether Turkey would allow the United States to use bases in that country in an operation against Iraq is unclear. Of particular concern to Turkey is that upheaval in Iraq could lead to the creation of an independent Kurdish state in northern Iraq. This, in turn, could energize Kurds in neighboring Turkey to seek independence as well. The INC has received no training or equipment, which upsets members of Congress, some of whom recently implored President Bush to start military training for the INC and to provide them with funding support as well.

Iraq has also received recent international attention as a state that is hostile toward minority religions. Although Shi'a Arabs are the largest religious group in Iraq, Sunni Arabs dominate the economic and political life. The Iraqi government systematically discriminates against the Shi'as, severely restricting or banning many Shi'a religious practices. For decades, the regime has conducted a campaign of murder, execution, arrest, and extended detention against Shi'a religious leaders and followers. The Iraqi government often destroys or desecrates Shi'a mosques and holy sites, disrupts religious ceremonies, and interferes with Shi'a religious education. It has recently banned the broadcasts of Shi'a religious programs on government-controlled radio and television, and the publication of Shi'a books. Saddam Hussein's regime consistently interferes with religious pilgrimages, both of Iraqi Muslims who desire to travel to the holy cities of Mecca and Medina in Saudi Arabia and of the Iraqi and non-Iraqi Muslim pilgrims who want to travel to holy sites within Iraq.

It appears that Iraq will never realize its full potential while Saddam Hussein is in power.

# Facts at a Glance

**Background** — Formerly part of the Ottoman Empire, Iraq became an independent kingdom in 1932. A republic was proclaimed in 1958, but in actuality a series of military strongmen has ruled the country since then, the latest being Saddam Hussein.

## Geography

**Location** — Middle East, bordering the Persian Gulf, between Iran and Kuwait; slightly more than twice the size of Idaho

**Land Boundaries** — Total: 2,256 mi (3,631 km); border countries: Iran 906 mi (1,458 km), Jordan 112 mi (181 km), Kuwait 150 mi (242 km), Saudi Arabia 505 mi (814 km), Syria 376 mi (605 km), Turkey 206 mi (331 km)

**Climate** — Mostly desert; mild to cool winters with hot, dry, cloudless summers; northern mountainous regions along Iranian and Turkish borders experience cold winters with occasionally heavy snows that melt early in the spring and sometimes cause extensive floods in central and southern Iraq

**Terrain** — Mostly broad plains; marshes along the Iranian border in the south with large flooded areas; mountains along the borders with Iran and Turkey

**Natural Resources** — Petroleum, natural gas, sulfur, phosphate

**Environment/Current Issues** — Government water control projects have drained most of the inhabited marsh areas east of An Nasiriyah by drying up or diverting the feeder streams and rivers. The destruction of the natural habitat poses serious threats to the area's wildlife populations, and there are inadequate supplies of potable (drinkable) water.

## People

**Population** — 23.6 million (July 2001 estimate)

**Ethnic Groups** — Arab 75–80%, Kurdish 15–20%, Turkman, Assyrian, or other 5%

**Religions** — Muslim 97% (Shi'a 60-65%, Sunni 32-37%), Christian or other 3%

**Languages** — Arabic, Kurdish (official language in the Kurdish region), Assyrian, Armenian

# Facts at a Glance

### Government

| | |
|---|---|
| **Government Type** | Republic; capital: Baghdad |
| **Administrative Divisions** | 18 provinces |
| **Legal System** | Based on Islamic law in special religious courts, civil law system elsewhere; has not accepted compulsory International Court of Justice jurisdiction |
| **Executive Branch** | Chief of state: President Saddam Hussein (since 16 July 1979) |
| **Legislative Branch** | Unicameral (one legislative chamber) National Assembly (250 seats; serve four-year terms) |
| **Judicial Branch** | Court of Cassation |
| **Political Parties and Leaders** | Ba'th Party (Saddam Hussein, central party leader) |

### Economy

| | |
|---|---|
| **Industries** | Petroleum, chemicals, textiles, construction materials, food processing |
| **Agriculture Products** | Wheat, barley, rice, vegetables, dates, cotton, cattle, sheep |
| **Export Commodities** | Crude oil; export partners: Russia, France, Switzerland, China (2000) |
| **Import Commodities** | Food, medicine, manufactures; import partners: Egypt, Russia, France, Vietnam (2000) |
| **Currency** | Iraqi dinar (IQD) |

(*Source:* CIA Factbook)

# History at a Glance

| | |
|---|---|
| **4000 B.C.** | The city of Sumer is built at the mouth of the Euphrates River on the Persian Gulf. It is the model for all the later city-states of the Middle East and the Mediterranean world. |
| **3000 B.C.** | Centered around their original city state of Ashur in northern Iraq, the Assyrians first emerge as a regional power. |
| **1700 B.C.** | Under Hammurabi, the city-state of Babylon (near modern Baghdad) first comes to prominence. Hammurabi writes the first legal code. |
| **1116 B.C.** | The Assyrian empire revives and reaches its greatest extent. |
| **616 B.C.** | Assyrian power finally ends and a new Babylonian Empire rises. |
| **661 A.D.** | The Islamic world falls under the leadership of the Umayyad Dynasty. Their capital and center of interest is in Damascus, but they control most of Iraq. |
| **763** | Baghdad is established as the new capital of the Abbasid Dynasty. |
| **1055** | Seljuk Turks conquer Baghdad. |
| **1258** | The Mongols destroy Baghdad. |
| **1533–1547** | Iraq becomes part of the Ottoman Empire. |
| **1918** | British troops drive remaining Turkish troops from Iraq, taking possession of the oil fields around Mosul. |
| **1955** | Iran, Iraq, Turkey, and Great Britain form a mutual defense organization called the Baghdad Pact. |
| **1958** | King Faisal is killed in a coup, and the Hashemite kingdom is replaced by a military government with Abdel Karim Kassem as the new prime minister. |
| **1959** | Iraq withdraws from the Baghdad Pact. |
| **1961** | The Kurdish minority concentrated in northwestern Iraq starts a revolt. |
| **1963** | Members of the Ba'thist party come to power. |
| **1970** | Iraq nationalizes oil deposits, most of which are owned by British companies. Britain attempts to freeze Iraqi bank deposits in London, but finds they have been moved to Switzerland. |

# History at a Glance

**1979** Saddam Hussein emerges as the dominant figure among the Ba'thist leadership.

**1990** Disputes over war debts and oil deposits lead to an Iraqi occupation of Kuwait (August 4).

**1991** Iraq is devastated by a massive military attack by an international coalition (January and February). Kuwait is liberated and some outlying Kurdish and Shi'ite areas of Iraq are occupied by Western troops.

**1990s** Saddam Hussein and the Ba'thist regime remain in power. Iraq frequently interferes with United Nations weapons inspectors looking for illegal weapons, but international support for another military attack is lacking.

**2000** The United States Department of State places Iraq on its list of nations that sponsor terrorism.

**2001** The terrorist attacks of September 11 on the United States bring Iraq under scrutiny for its possible involvement in with the terrorists and place the country on the list of potential future targets in the war on terrorism.

## Further Reading

Bratvold, Gretchen. *Iraq in Pictures.* Minneapolis, Minnesota: Lerner Publishing, 1992.

Britton, Tamara L. *Iraq.* Edina, Minnesota: ABDO Publishing Company, 2000.

Foster, Leila M. *Iraq.* New York: Scholastic Library Publishing, 1998.

Foster, Leila M. *The Persian Gulf War.* New York: Scholastic Library Publishing, 1991.

King, John. *A Family from Iraq.* Milwaukee: Raintree Publishers, 1997.

Nardo, Don. *The War Against Iraq.* Farmington Hills, Minnesota: Lucent Books, 2001.

Service, Pamela F. *Mesopotamia: Cultures of the Past.* Ipswich, Massachusettes: Marshall Vacendish Corporation, 1999.

Spencer, William. *Iraq: Old Land, New Nation in Conflict.* Brechenridge, Colorado: Twenty First Century Books, 2000.

Walner, Rosemary, and Paul J. Deegan. *Saddam Hussein: War in the Gulf.* Edina, Minnesota: ABDO and Daughters, 1991.

Whitcraft, Melissa. *The Tigris Euphrates Rivers.* New York: Scholastic Library Publishing, 1999.

# Bibliography

Ahmed, Akbar S. *Discovering Islam: Making Sense of Muslim History and Society.* Boston: Routledge & Kegan Paul, 1987.

Arnov, Anthony and Ali Abunimah. *Iraq under Siege: The Deadly Impact of Sanctions and War.* Cambridge, MA: South End Press, 2000.

Bradshaw, Michael. *A World Regional Geography: The New Global Order.* Madison, WI: Brown and Benchmark Publishers, 1999.

Central Intelligence Agency. *CIA—The World Fact Book, Iraq.* www.cia.gov/cia/publications/factbook (current).

De Blij, H.J. and Peter O. Muller. *Geography: Realms, Regions, and Concepts.* New York: John Wiley and Sons, 1999.

Fisher, William B. *The Middle East: A Physical, Social, and Regional Geography.* London: Methuen, 1978.

Fromkin, David. *A Peace To End All Peace: Creating the Modern Middle East, 1914–1922.* New York: Henry Holt, 1989.

Gilsenan, Michael. *Recognizing Islam: Religion and Society in the Middle East.* London: I.B. Tauris, 1992.

Hourani, Albert, et al. *The Modern Middle East: A Reader.* Berkeley, CA: University of California Press, 1994.

Kreyenbroek, Philip G. *The Kurds: A Contemporary Overview.* London: Routledge, 1992.

Lewis, Bernard. *The Middle East: A Brief History of the Last 2,000 Years.* New York: Scribner, 1996.

Longrigg, Stephen H. *The Middle East: A Social Geography.* Chicago: Aldine Press, 1970.

Meiselas, Susan, Martin Van Bruimessan and A. Whitley. *Kurdistan: In the Shadows of History.* New York: Random House, 1997.

United States Department of State—*Background Notes on Iraq.* http://www.state.gov/r/pa/ei/bgn/6804.htm (current).

Wright, Robin B. *Sacred Rage: The Crusade of Modern Islam.* New York: Linden Press, 1985.

# Index

Abbadid caliphate, 30
Agriculture, 11, 14, 18, 20, 21, 23, 24, 26, 69, 74
Airports, 81
Al Basrah, 31, 32, 35, 38, 81
Alexander the Great, 29
Al Hijaz, 33
Allen, Edmund, 32
Anglo-Persian Oil Company, 32, 36
Arab conquest, 30
Arabian horse, 74
Arab-Israeli War of 1973, 43
Arab League, 38-39, 43, 67
Arab nationalism, 32
Arab Union of Jordan and Iraq, 41
Arab World, 9-10
Area, 17
Arif, Abdul Rahman, 42
Arif, Abdul Salam, 42
Armenians, 51, 56
Assyria, 10, 29
Assyrians, 51, 55-56, 59

Babylon, 10, 29
Baghdad, 10, 13, 23, 30-31, 35, 56, 74, 81
Baghdad Pact, 40, 41
Bakr, Ahmed Hassan al-, 42, 44, 62
Baranzi, Masoud, 62
Barzani, Mustafa al-, 44
Basra, 13, 81
Ba'th Party, 13, 42, 61, 62, 65, 66
Bedouins, 54-55
Birds, 26
Boundaries, 11, 13, 18
Britain
    and Iraq mandate, 34, 35
    and Kuwait, 41-42
    and modern state of Iraq, 10, 34-36, 61
    and oil, 32
    and Ottoman Turks, 32-34
    and rule over Iraq, 32-34, 38
    and treaty of alliance and protection, 35-36, 38
    and war with Iraq, 38
    and World War I, 32-34
Bush, George H.W., 47, 84, 86
Buwayhids, 30

Central Treaty Organization, 41
Christian Nestorian Church, 56
Christians, 59
Cities, 10, 13, 21, 51, 77
Climate, 21, 22-24
Coastline, 11
Communist Party, 85
Constitution, 63-64
Coups
    of 1936, 37-38
    of 1958, 41, 61
    of 1963, 42
    of 1968, 42

Daily life, 77-81
Dates, 21, 74
Da'wa Islamic Party, 62
Delta, 21
Desert, 17-18, 22
Dictatorship, 13
Dinar, 48, 73, 75
Dissidents, 84-85
Dust storms, 24

Economy, 14-15, 48, 69-75, 78, 83
    and sanctions, 14, 45, 47, 48, 69, 70, 81, 83, 85
Education, 77, 78, 79-80
Eisenhower Doctrine, 41
Elections, 40, 62, 65
Environmental issues, 26
Ethnic groups, 51-56
Euphrates River, 11, 17, 18, 20, 21, 24, 26, 74
Exports, 70, 71, 73-74

Faisal I, 33, 34-35, 36
Faisal II, 38, 40, 41
Family, 78
Fish, 26-27, 74

# Index

Flag, 59
Floods, 18, 20, 21, 24, 26
Food, 72
    and oil-for-food program, 14-15, 48, 71
Forests, 20
Future, 83-87

Gailani, Rashid Ali al-, 38
Germany, 32, 38
Ghazi, King, 36, 37, 38
Government, 13
    and dictatorship, 61-67
    and Iraq as independent state, 36-38
    and monarchy, 33, 34-38, 40, 41, 61
    and new state of Iraq, 34-36
    and Ottoman Turks, 31-32
    pro-Western, pan-Arab, 40
    and republic, 41-42, 61
Grasslands, 17, 18, 20
Greeks, 29
Gross domestic product, 70, 71

Health standards, 78-79
Highlands, 17, 20
History, 10, 18, 29-49
Hulagu, 30
Hussein I, King of Jordan, 41
Hussein, Saddam, 13, 15, 77
    and economy, 70, 71, 83
    and Kurds, 14, 45-48, 66
    and Kuwait's independence, 47
    and Marsh Arabs, 21, 53
    opposition to, 86-87
    as president, 44, 61, 62-63, 65
    and religion, 87
    and weapons inspections, 14, 48-49

Imports, 73-74
Independence, 36-38
Industry, 69, 74
Infrastructure, 14, 20, 69, 72, 80-81
Iran, 11
    and Iran-Iraq War, 11, 13, 14, 18, 26, 44-45, 52, 69, 70
    and Islamic republic, 44
    and Kurds, 14, 44, 46, 52
    and Persians, 29, 31
Iran-Iraq War, 11, 13, 14, 18, 26, 44-45, 52, 69, 70
"Iraq," meaning of, 17
Iraqi Intelligence Service, 84-85
Iraqi National Congress, 86
Iraq Petroleum Company, 36, 73
Irrigation, 11, 21, 23, 30
Islam, 56-59, 63
    and education, 80
    and religious courts, 66
    Shi'a, 13, 30, 35, 44, 46, 47, 53, 56-57, 58, 62, 77, 87
    spread of, 10, 30
    Sunni, 30, 35, 44, 56-57, 58, 77, 87
Israel, 42-43, 67
    and Arab-Israeli War of 1973, 43
    and attack on nuclear reactor, 44
    and Hussein, 15
    and Jordan, 39, 40
    and 1948 war, 39
    and Persian Gulf War, 15
    and Six-Day War, 42
    and Suez Crisis, 40-41

Jews, 51, 59
Jordan
    Iraqi troops in, 41
    Iraq joined with, 41
    and Israel, 39, 40
    and 1991 Gulf War, 11
    and Palestinians, 43
Judicial system, 65-66

Kassem, Abdul Karim, 41, 42
Kirkuk, 14, 20, 52, 70
Kurdistan Democratic Party, 47-48, 62
Kurds, 13-14, 31, 51, 52-53, 57, 63
    and civil war, 47-48
    and government, 66-67
    and Hussein, 14, 45-48, 66
    and independent state, 86-87
    and new state of Iraq, 35

# Index

and peace with government, 42
and political parties, 47-48, 62, 67
and protection of after Persian Gulf War, 14, 46-47, 66
and revolts against government, 14, 39, 42, 44, 45-46, 77
Kuwait, 11
   and Britain, 41-42
   Iraqi claim on, 42
   and Persian Gulf War, 11, 13, 14, 15, 21, 26, 45-47, 52, 69, 70-71, 81, 83
   recognition of independence of, 47

Languages, 51, 52, 56, 59
Lawrence, T.E., 33
League of Nations, 34, 36
Livestock, 74

Mamluks, 30, 31
Mammals, 26-27
Manufacturing, 69, 74
Marsh Arabs, 21, 53
Ma'ruf, Taha Muhyi al-Din, 65
Mesopotamia, 17, 29, 59
Middle East Treaty Organization, 40
Midhat Pasha, 31
Military, 36, 37-38, 41, 45, 61, 71
Minerals, 70
Monarchy, 33, 34-38, 40, 41, 61
Monetary system, 48, 73, 75
Mongols, 30-31, 56
Mosul, 31, 35, 36, 52, 70
Mountains, 20, 24
Muhammad, 10, 58

National Assembly, 63-64
Natural landscapes, 17-18, 20-25, 26-27
1991 Gulf War, 11

Oil, 9, 11, 13, 14, 20, 69, 70, 77
   and Britain, 32
   development of, 36
   and environment, 26
   illegal trade in, 15, 71
   and Kurds, 14
   nationalization of, 73
   and oil-for-food program, 14-15, 48, 71
   and Persian Gulf War, 13, 45
   profits from, 40, 71, 73
   and republic, 41
   and sanctions, 26
   and Six-Day War, 42
   and Soviet Union, 39
Opposition parties, 47-48, 62, 67
Organization of Petroleum Exporting Countries, 13, 67, 73
Organization of the Islamic Conference, 67
Ottoman Turks, 31-34

Pan-Arabism, 35, 36, 37, 40, 63
Parliament, 36
Patriotic Union of Kurdistan, 47, 48, 62
People, 13-14, 51-56, 63
Per capita income, 71
Persian Gulf War, 11, 13, 14, 15, 21, 26, 45-47, 52, 69, 70-71, 81, 83
   and sanctions, 14, 45, 47, 48, 69, 70, 78, 81, 83, 85
Persians, 29, 31
   as ethnic group, 51, 56
Plains, 17, 20-21, 22
Population, 11, 13-14, 51-52, 70
Ports, 11, 81
Poverty, 71-73, 77
President, 61, 62, 64, 65
   *See also* Hussein, Saddam
Provinces, 66-67

Qadisiyya, Battle of, 30

Radio Free Europe/Radio Liberty, 84
Railroad, 80
Rainfall, 2, 18, 23
Ramadan, Taha Yasin, 65
Religion, 51, 56, 59, 63, 87
   *See also* Islam
Reptiles, 27

# Index

Republic, 13, 41-42, 61
Revolutionary Command Council, 42, 61-62, 63-64, 65
Rivers. *See* Euphrates River; Tigris River
Roads, 20, 69, 80-81
Rural areas, 51-52, 77-78

Sabeans (Christians of St. John), 51, 56
Said, Nuri as-, 38, 39, 41
Sassanian Empire, 30
Saudi Arabia, 11, 37, 85
Seleucids, 29
Seljuks, 30
Service industry, 74
Shatt-al-Arab, 11, 13, 18, 44
Six-Day War, 42
Social services, 78-79
Soil, 24, 26
Soviet Union, 38, 39, 42
Suez Crisis, 40-41
Sumer, 10, 29
Supreme Assembly of the Islamic Revolution in Iraq, 62
Syria, 11, 43
Syrian Desert, 17-18

Talabani, Jalal, 62
Tamerlane, 30-31, 56
Terrorism, 15, 83-86
Tigris River, 11, 17, 18, 20, 21, 24, 26, 32, 74, 81
Trade, 69, 73-74
　and sanctions, 14, 45, 47, 48, 69, 70, 78, 81, 83, 85
　*See also* Oil
Transjordan, 39
　*See also* Jordan
Transportation, 20, 69, 80-81
Tribes, 52, 53, 54-55, 74, 78
Turkey, 11
　and Kurds, 14, 46, 52, 53, 86-87
　and mutual defense treaty, 40
　and opposition to Hussein, 86-87
Turkomans, 51, 56

Umm Qasr, 11, 81
United Arab Republic, 41
United Nations
　Iraq as member of, 67
　and Iraqi claim to Kuwait, 42
　and Kurds, 14
　and 1991 Gulf War, 11
　and oil-for-food program, 14-15, 48, 71
　and Persian Gulf War, 14, 45
　and sanctions, 14, 26, 45, 47, 48, 69, 70, 78, 81, 83, 85
　and Suez Crisis, 40-41
　and weapons inspections, 14, 47, 48-49
United States
　and aid to military, 45
　and air strikes, 48-49
　and cruise missile attack, 47
　and Eisenhower Doctrine, 41
　and Iraq as future target in war on terrorism, 15, 83-84, 86
　and Kurds, 46-47, 48, 67
　and opposition to Hussein, 86-87
　and Persian Gulf War, 45, 81
　and Six-Day War, 42

Vegetation, 18, 20-21

Water, 9, 11, 70, 77
Weapons inspections, 14, 47, 48-49
Wildlife, 26-27
Winds, 21, 24
World War I, 10, 32-33
World War II, 38-39

Yazidis, 51, 56

## Picture Credits

page:

| | | | |
|---|---|---|---|
| 8: | Grant Smith/Corbis | 50: | Roger Wood/Corbis |
| 12: | 21st Century Publishing | 54: | Nik Wheeler/Corbis |
| 15: | AP/Wide World Photos | 55: | Dave Bartruff/Corbis |
| 16: | Corbis | 57: | AFP/Corbis |
| 19: | 21st Century Publishing | 60: | AP/Wide World Photos |
| 22: | Charles & Josette Lanars/Corbis | 63: | AFP/Corbis |
| 25: | David Lees/Corbis | 64: | AP/Wide World Photos |
| 28: | David Lees/Corbis | 68: | AP/Wide World Photos |
| 33: | Hulton-Deutsch Collection/Corbis | 72: | AP/Wide World Photos |
| 37: | Hulton-Deutsch Collection/Corbis | 75: | AFP/Corbis |
| 43: | Bettmann/Corbis | 76: | AP/Wide World Photos |
| 46: | Peter Turnley/Corbis | 79: | AP/Wide World Photos |
| 49: | AP/Wide World Photos | 82: | AP/Wide World Photos |

Frontis: Flag courtesy of *theodora.com/flags*. Used with permission.

Cover: © Francois de Mulder/Corbis

## About the Author

**ANGELIA L. MANCE** is the Associate Director for the National Council for Geographic Education and the Coordinator for the National Geographic Society's Alabama State Geographic Bee. Angelia has taught geography at the college level for the past 8 years. Her professional interests include world regions, religions, and cultures.

**CHARLES F. "FRITZ" GRITZNER** is Distinguished Professor of Geography at South Dakota State University. He is now in his fifth decade of college teaching and research. Much of his career work has focused on geographic education. Fritz has served as both president and executive director of the National Council for Geographic Education and has received the Council's George J. Miller Award for Distinguished Service.